SpringerBriefs in Computer Science

T0183772

More information about this series at http://www.springer.com/series/10028

Lidija Čomić • Leila De Floriani
Paola Magillo • Federico Iuricich

Morphological Modeling of Terrains and Volume Data

 Springer

Lidija Čomić
Department for Fundamental Disciplines
Faculty of Technical Sciences
University of Novi Sad
Novi Sad, Serbia

Paola Magillo
Department of Computer Science
  Bioengineering, Robotics
  and Systems Engineering
University of Genova
Genova, Italy

Leila De Floriani
Department of Computer Science
  Bioengineering, Robotics
  and Systems Engineering
University of Genova
Genova, Italy

Federico Iuricich
Department of Computer Science
University of Maryland
College Park, MD, USA

ISSN 2191-5768   ISSN 2191-5776 (electronic)
ISBN 978-1-4939-2148-5   ISBN 978-1-4939-2149-2 (eBook)
DOI 10.1007/978-1-4939-2149-2
Springer New York Heidelberg Dordrecht London

Library of Congress Control Number: 2014951706

Printed on acid-free paper

Springer is part of Springer Science+Business Media (www.springer.com)

Preface

Scalar fields are real-valued functions defined point-wise within a d-dimensional domain. They appear in many applications, including physics, chemistry, medicine, geography, etc., and represent real or simulated phenomena characterized by a spatial extension. The most common example are height fields which describe terrains and have been studied extensively in Geographic Information Systems (GISs) and scientific visualization. Scalar fields can also be defined on shapes (which describe the surface of an object in 3D) to represent some point-wise defined property on them (such as curvature).

As d-dimensional domains are discretized (e.g., as d-dimensional images composed of voxels) and so are shapes (e.g., a surface tessellated as a mesh of triangles), scalar fields defined on them have a discrete representation: usually, field values are associated with the voxels of an image or with the vertices of a mesh. This provides a detailed representation of the scalar field.

Thanks to hardware and software development, available data representing both the field and its domain are increasing in size and complexity. On the other hand, a detailed representation is verbose and not suitable to analysis tasks such as recognition and classification. Therefore, the issue arises to switch from representation to description of a scalar field. While a representation provides all details necessary to know the field point-wise, a description is more abstract and has the purpose of showing the main characteristics of the field, such as, for instance, its maxima/minima/saddles and their relative positions.

This book focuses on morphological descriptions of scalar fields, mainly in 2D (height fields or terrains) and 3D (volume data). Specifically, we consider morphological descriptions based on identifying maxima, minima and saddles, finding their influence zones, and encoding their mutual spatial relations. All this is formalized through Morse theory, Morse and Morse-Smale complexes. We provide the mathematical background, which has been formally defined for smooth functions and then transposed into a discrete setting in different ways. We introduce the main algorithmic approaches and their characteristics, and present algorithms for discrete scalar fields in two and three dimensions, and (where possible) in general dimensions.

v

Although morphological descriptions based on Morse and Morse-Smale complexes represent scalar fields in a much more compact way than the initial geometric representation, simplification of these complexes is often necessary. Simplification allows disregarding less meaningful morphological features, or spurious features, due to noise in the data, as well as adapting the level of abstraction of the description to the current application task. Therefore, we present simplification operators which act on the morphological description of a scalar field.

The issue of simplification comes along with that of multi-resolution, which requires being able to build a compact model encompassing different levels of detail (corresponding to different degrees of simplification), in such a way that the appropriate level of detail can be extracted on the fly according to user-defined criteria.

This book is organized as follows. Chapter 1 contains the necessary mathematical background on Morse theory in the smooth and in the discrete case. Chapter 2 presents a classification of existing algorithms for morphological computation based on different and often orthogonal criteria, where the main criterion is the algorithmic approach they use. Such a criterion identifies boundary-based and region-growing methods (which essentially deal with 2D and 3D scalar fields), watershed-based and methods based on a discrete Morse theory due to Forman (which are dimension-independent). The next three chapters present a survey of algorithms belonging to the first two classes (Chap. 3), of watershed-based algorithms (Chap. 4) and of Forman-based algorithms (Chap. 5). Chapter 6 considers the issues related to simplification and multi-resolution. Finally, Chap. 7 presents experimental comparisons and draws concluding remarks.

Novi Sad, Serbia   Lidija Čomić
Genova, Italy   Leila De Floriani
Genova, Italy   Paola Magillo
College Park, MD, USA   Federico Iuricich
July 2014

Acknowledgements

We wish to thank all colleagues and collaborators who have worked with us on morphology of terrains and scalar fields over the years. Specifically, we want to thank former Ph.D. students at the University of Genova and at the University of Maryland and current collaborators, Emanuele Danovaro, Riccardo Fellegara, Maria Vitali and Kenneth Weiss, who have extensively worked on the project of morphological shape analysis. We would like to acknowledge the support of Mostefa Mesmoudi and Laura Papaleo who contributed to several of our papers on this subject. Last, but not least, a special thanks to our colleague Enrico Puppo for the many stimulating discussions on the topic of the book.

We would like to thank all the people from Springer-Verlag who were involved in the publication of this book. We are particularly grateful to Susan Lagerstrom-Fife and Jennifer Malat for their constant help and encouragement through the development of this project.

We want to thank the funding agencies for their support. This work has been partially supported by the US National Science Foundation under grant number IIS-1116747, and by the Italian Ministry of Education and Research under the PRIN 2009 program, within the project 2009MT4K2S.

All terrain and 3D surface datasets used in this work are courtesy of the Virtual Terrain Project (VTP) at http://vterrain.org/, of the USGS National Elevation Dataset at http://ned.usgs.gov/, and of the AIM@Shape repository at http://www.aimatshape.net/. All volume datasets are courtesy of the Volume Library at http://www9.informatik.uni-erlangen.de/External/vollib/.

Contents

Chapter 1
Background

In this chapter, we introduce the mathematical structures used to represent scalar fields and their morphology in the smooth and in the discrete cases.

In Sect. 1.1, we provide some mathematical background. We introduce the concept of manifold, which will be used to characterize the domain of scalar fields, and the concept of cell complex with its special cases (regular grid, and simplicial complex). In Sect. 1.2, we introduce models for scalar fields in a real-world setting, where the function is known only at a set of sampled points, such as regular models and simplicial models.

In Sect. 1.3, we present the basic notions of Morse theory [17,19], which provides a description of the morphology of functions and their domains in the smooth case. In Sect. 1.4, we present the watershed transform in the smooth case, which is an independently developed framework, alternative to Morse theory.

Two approaches exist for extending the results of Morse theory to the discrete case, as required when dealing with scalar field models based on regular grids or simplicial complexes [6]. Banchoff's piecewise-linear Morse theory [1,2], presented in Sect. 1.5, transposes the results obtained on smooth functions to the case of a function having values at the vertices of a complex with polygonal cells, while elsewhere the function is approximated through linear interpolation. This theory has been defined in 2D, and then extended to the 3D case. Forman's discrete Morse theory [11], presented in Sect. 1.6, extends Morse theory to the discrete case where a function value is defined on all cells of a cell complex. This theory is entirely combinatorial, and completely dimension-independent.

1.1 Some Preliminary Definitions

Here, we introduce the notions of manifold [16, 26], cell complex [15], simplicial complex [26], and regular grid [6].

© The Author(s) 2014
L. Čomić et al., *Morphological Modeling of Terrains and Volume Data*,
SpringerBriefs in Computer Science, DOI 10.1007/978-1-4939-2149-2_1

1.1.1 Manifolds

The *closed d-dimensional ball* is the set $\{p = (x_1, \ldots, x_d) \in \mathbb{R}^d \mid ||p|| \leq 1\}$, and the *open d-dimensional ball* is the set $\{p = (x_1, \ldots, x_d) \in \mathbb{R}^d \mid ||p|| < 1\}$, where $|| \cdot ||$ denotes the norm of the point viewed as a vector (i.e., its Euclidean distance from the origin). A *(closed or open) half d-dimensional ball* is the intersection of the (closed or open) d-dimensional ball with the half-space $\{p = (x_1, \ldots, x_d) \in \mathbb{R}^d \mid x_1 \geq 0\}$.

Intuitively, a *d-dimensional manifold* is a subset of $\mathbb{R}^n$ which is locally d-dimensional at each point. This concept is formalized in the following definition.

Definition 1. A subset M of $\mathbb{R}^n$ is a *d-manifold* if each point $p \in M$ has a neighborhood which is homeomorphic to the open d-dimensional ball.

A subset M of $\mathbb{R}^n$ is a *d-manifold with boundary* if each point $p \in M$ has a neighborhood which is homeomorphic either to the open d-dimensional ball, or to the open half d-dimensional ball.

Intuitively, two sets are homeomorphic if they are *topologically equivalent*: each can be transformed into another by just deforming, without tearing or cutting [12].

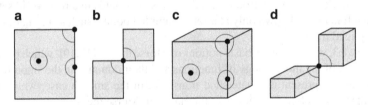

Fig. 1.1 Some sets which satisfy, or do not satisfy, the definition of a 2-manifold. Note that a two-dimensional ball is a disk. Set (**a**) in 2D is a 2-manifold with boundary, set (**b**) is not a manifold. Set (**c**) in 3D is a 2-manifold without boundary, set (**d**) is not a manifold

The set $\mathbb{R}^n$ and the open n-dimensional ball are n-manifolds (without boundary), while the closed n-dimensional ball is an n-manifold with boundary. If we restrict our attention to bounded and closed subsets of $\mathbb{R}^n$, we have that only d-manifolds, with $d < n$, can be without boundary in $\mathbb{R}^n$. For example, the surface bounding a solid in $\mathbb{R}^3$ is a 2-manifold without boundary. In $\mathbb{R}^n$, an n-manifold (which is a closed bounded set) must have a boundary. An example is a disc (filled circle) in $\mathbb{R}^2$ or a ball (filled sphere) in $\mathbb{R}^3$. Figure 1.1 shows some examples of manifold and non-manifold sets, for $d = 2$.

The *Euler characteristic* of a d-dimensional manifold M is defined as

$$\chi(M) = \beta_0 - \beta_1 + \beta_2 - \beta_3 + \cdots + (-1)^d \beta_d = \sum_{i=0}^{d} (-1)^i \, \beta_i \qquad (1.1)$$

where β_i is the i-th Betti number of M.

Intuitively, β_0 counts the number of connected components of an object; β_1 counts the numbers of through holes of an object. Betti number $\beta_2 = 0$ for 2-manifolds in $\mathbb{R}^2$, while, for 2-manifolds and 3-manifolds in $\mathbb{R}^3$, β_2 counts the numbers of cavities (voids) inside an object. In $\mathbb{R}^3$, all other Betti numbers are null.

1.1.2 Cell Complexes

In this book, we will consider scalar fields as scalar functions defined over manifold subsets of $\mathbb{R}^n$ (mainly for $n = 2$ and $n = 3$). For computational purposes, the manifold domain of a scalar field will be represented by a decomposition into a collection of elements with simple shapes. Such basic elements are formalized as cells, and their collection is formalized through the notion of cell complex.

Definition 2. A *d-dimensional cell* in $\mathbb{R}^n$, or a *d-cell*, is a subset γ of $\mathbb{R}^n$ which is homeomorphic to the closed d-dimensional ball; d is called the *dimension* of γ.

The (relative) interior $i(\gamma)$ of a d-cell γ is the image of the open ball under the same homeomorphism. Its (relative) boundary is $b(\gamma) = \gamma \setminus i(\gamma)$.

Intuitively, a *cell complex* in $\mathbb{R}^n$ is a finite collection Γ of cells, of different dimensions, which are glued together in a consistent way: the interiors of the cells are disjoint, and the boundary of each cell is a union of interiors of (lower-dimensional) cells belonging to Γ.

Definition 3. A d-dimensional *cell complex* (a cell d-complex) Γ is a collection of cells, such that:

- $i(\gamma_1) \cap i(\gamma_2) = \emptyset$ for each pair of distinct cells γ_1 and $\gamma_2 \in \Gamma$;
- the boundary of each cell in Γ is a disjoint union of interiors of cells of Γ;
- the maximum dimension of cells in Γ is d.

The *domain* of a cell complex Γ is the point set in $\mathbb{R}^n$ given by the union of all cells of Γ.

In a cell complex Γ, a cell γ' is a *face* of another cell γ, if $\gamma' \subseteq \gamma$; it is a *proper face* if $\gamma' \subset \gamma$. The proper faces of a cell γ have a lower dimension than that of γ. If γ' is a face of γ, then γ is a *co-face* of γ'. Note that the proper faces of a cell are not defined in relation with the cell itself, but only in relation with a cell complex containing it.

The *(combinatorial) boundary* of a cell γ in a cell complex Γ is the set of the proper faces of γ (see Fig. 1.2a). The *(combinatorial) co-boundary*, or *star* of γ, denoted $*\gamma$, is the set of the co-faces of γ, i.e., the set of cells of Γ which are incident in γ (see Fig. 1.2b). The *link* of γ, denoted $Lk(\gamma)$ contains those faces of cells of $*\gamma$, which are not in $*\gamma$ (see Fig. 1.2c).

A cell γ of a d-dimensional cell complex Γ is *maximal* if its dimension is d. A d-dimensional cell complex Γ is *pure* if any k-dimensional cell, with $k < d$, is a face of a d-dimensional cell. Intuitively, this implies that the domain of the cell complex is uniformly d-dimensional, without dangling parts of lower dimension.

 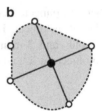 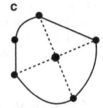

Fig. 1.2 (**a**) The boundary of the 2-cell (*gray*) is composed of four 1-cells (*bold lines*) and four 0-cells (*full dots*). (**b**) The star of the central vertex is composed of four 1-cells (*bold lines*) and four 2-cells (*gray surfaces*) incident in it. (**c**) The link of the central vertex is composed of the five 1-cells (*bold lines*) in the neighborhood of the vertex as well as the five 0-cells (*full dots*) incident in them

Cell complexes used in practice to describe the domain of a d-dimensional scalar field are pure, and have a manifold domain. Thus, in the following we will implicitly assume cell complexes with manifold domain.

The *Euler characteristic* of a d-dimensional cell complex Γ is defined as

$$\chi(\Gamma) = m_0 - m_1 + m_2 - m_3 + \cdots + (-1)^d m_d = \sum_{i=0}^{d} (-1)^i \, m_i \qquad (1.2)$$

where m_i denotes the number of cells of dimension i in the complex. The Euler characteristic of a cell complex (with manifold domain) is equal to the Euler characteristic of its domain.

Definition 4. The *k-skeleton* of a d-dimensional complex Γ (with $k \leq d$) is the subset of Γ containing the cells of Γ of dimension less than or equal to k.

The k-skeleton is a k-dimensional cell complex. Specifically, the *1-skeleton* is made up of cells of dimension 0 (vertices) and 1 (edges). Disregarding the geometry of the cells, the 1-skeleton has the combinatorial structure of a graph $G = (V, E)$, where the nodes in V correspond to the vertices of Γ and the arcs in E correspond to the edges of Γ (where each edge is seen as the pair of its endpoint vertices).

1.1.3 Regular Grids and Simplicial Complexes

Although theoretically cells in $\mathbb{R}^n$ can be of any shape that is topologically equivalent to a ball, cells used in practice have some standard, simple shape, i.e., are hyper-rectangles or simplices, and corresponding cell complexes are regular grids or simplicial complexes.

Definition 5. An axis-parallel d-dimensional hyper-rectangle ρ in $\mathbb{R}^n$ is the Cartesian product of n closed intervals, where exactly d of them are non-degenerate, i.e.,

$\rho = \{p = (x_1, \ldots, x_n) \in \mathbb{R}^n \mid x_i \in [a_i, b_i]\}$, where $\#\{i \mid a_i < b_i\} = d$. We say that ρ is *generated* by intervals $[a_i, b_i]$.

Examples of d-hyper-rectangles are a point $(d = 0)$, a straight-line segment $(d = 1)$, a rectangle $(d = 2)$, a cuboid $(d = 3)$. Usually, intervals have integer endpoints and unit length, i.e., $a_i \in \mathbb{Z}$, and $b_i = a_i$ or $b_i = a_i + 1$.

Definition 6. A d-*dimensional simplex*, or d-*simplex*, σ is the convex hull of $d + 1$ affinely independent points $p_0, \ldots, p_d$ in $\mathbb{R}^n$. Points $p_0, \ldots, p_d$ are called the *vertices* of σ.

Examples of simplices are a point (0-simplex), a straight-line segment (1-simplex), a triangle (2-simplex), a tetrahedron (3-simplex). Any d-dimensional hyper-rectangle, and any d-simplex, is also a d-dimensional cell.

A d-*dimensional regular grid* is a d-dimensional cell complex where all cells are hyper-rectangles, and the domain is also a d-dimensional hyper-rectangle. A 2D regular grid is also called a *square grid*, and a 3D regular grid is called a *cubic grid*.

Differently from generic cells, the proper faces of a hyper-rectangle ρ or a simplex σ are uniquely defined by the shape of ρ or σ, independently of the complex containing it. Given a hyper-rectangle ρ, generated by intervals $[a_i, b_i], i = 1, \ldots n$, any hyper-rectangle ρ' generated by intervals $[a_i', b_i']$, with either $a_i' = a_i$ and $b_i' = b_i$, or $a_i' = b_i' = a_i$, or $a_i' = b_i' = b_i$, is a *face* of ρ. Hyper-rectangle ρ' is a *proper face* of ρ if $\rho' \neq \rho$. This leads to an equivalent definition of regular grids.

Definition 7. A *regular grid* in $\mathbb{R}^n$ is a finite collection G of hyper-rectangles of different dimensions, such that:

- for any hyper-rectangle $\rho \in G$, all hyper-rectangles that are proper faces of ρ are in G;
- for any pair of hyper-rectangles $\rho_1, \rho_2 \in G$, either $\rho_1 \cap \rho_2 = \emptyset$, or $\rho_1 \cap \rho_2$ is a hyper-rectangle of G;

and the domain of G is a hyper-rectangle in $\mathbb{R}^n$.

Similarly, a d-*dimensional simplicial complex* is a d-dimensional cell complex where all cells are simplices. The following is an equivalent definition of a simplicial complex:

Definition 8. A *simplicial complex* in $\mathbb{R}^n$ is a finite collection Σ of simplices, of different dimensions, such that:

- for any simplex $\sigma \in \Sigma$, all simplices that are proper faces of σ are in Σ;
- for any pair of simplices $\sigma_1, \sigma_2 \in \Sigma$, either $\sigma_1 \cap \sigma_2 = \emptyset$, or $\sigma_1 \cap \sigma_2$ is a simplex of Σ.

The two definitions are equivalent because also the proper faces of a simplex σ are uniquely defined by the shape of σ, independently of the complex containing it. Given a simplex σ, any simplex σ', whose vertices are a (proper) subset of the vertices of σ, is a *(proper) face* of σ.

The interesting case for us is that in which the domain is a d-manifold. A 2D simplicial complex with manifold domain is also called a *triangle mesh*, and a 3D simplicial complex with manifold domain is also called a *tetrahedral mesh*.

A regular grid can always be transformed into a simplicial complex by decomposing each hyper-rectangle into simplices. In 2D, a regular grid is transformed into a triangulation by dividing each square cell into two triangles. In 3D, a regular grid is transformed into a tetrahedralization by dividing each cuboid into either five or six tetrahedra (see Fig. 1.3).

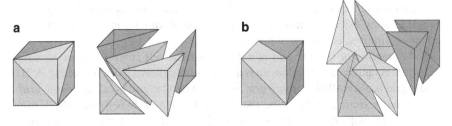

Fig. 1.3 Decomposition of a cube into (**a**) five and (**b**) six tetrahedra

1.1.4 Primal and Dual Complex

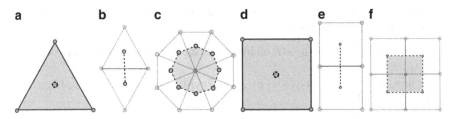

Fig. 1.4 The primal/dual relationships in a triangle mesh and in a square grid. (**a**) The dual of a triangle is a vertex (*dotted*); (**b**) the dual of an edge shared by two triangles is an edge (*dotted*); (**c**) the dual of a vertex is a polygon (*shaded*). (**d**) The dual of a square is a vertex (*dotted*); (**e**) the dual of an edge shared by two squares is an edge (*dotted*); (**f**) the dual of a vertex is a square (*shaded*)

Each d-dimensional cell complex Γ has a corresponding *dual* cell complex Γ^*, which is also a d-dimensional cell complex. With respect to its dual, Γ is called the *primal* cell complex. The 0-cells (vertices) of the dual complex Γ^* correspond to the d-cells of Γ, and can be geometrically placed at their centroids. The 1-cells of Γ^* correspond to the $(d - 1)$ cells of Γ, and so on. The maximal cells of Γ^* correspond to the vertices of Γ. In the following, we will call *primal* the cells of Γ and *dual* the cells of Γ^*.

When the primal cell complex Γ is a simplicial complex, the dual cell complex Γ^* is not simplicial since its cells may have an arbitrary shape (see Fig. 1.4a–c).

Each primal triangle corresponds to a dual vertex. Each primal edge, shared by two primal triangles, corresponds to a dual edge having the corresponding dual vertices in its boundary, and, similarly, a primal vertex corresponds to a dual 2-cell.

When the primal cell complex Γ is a regular grid, then Γ and Γ^* are both regular grids, as illustrated in Fig. 1.4d–f.

We call *primal graph* the graph representing the combinatorial structure of the 1-skeleton of the primal complex (see Sect. 1.1.2). Similarly, we call *dual graph* the graph representing the combinatorial structure of the 1-skeleton of the dual complex. Thus, given a cell complex Γ, the nodes of the primal graph correspond to the vertices of Γ and the arcs of the primal graph correspond to the edges of Γ (i.e., they represent the vertex-vertex adjacency relation in Γ). The nodes of the dual graph correspond to the d-cells of Γ, and its arcs correspond to pairs of d-cells which are mutually adjacent along a $(d-1)$-dimensional face in Γ.

1.2   Models for Scalar Fields

Here, we introduce scalar fields and their discrete models, based on cell complexes (for details, see [6]).

A d-dimensional scalar field is any phenomenon in the real or virtual world which is characterized by spatial extension within a d-dimensional domain $D \subseteq \mathbb{R}^n$ and by a function value existing in each point of the domain. Examples are temperature within a volume, pressure over a plane, or curvature over a surface embedded in 3D space. A typical case of a 2D scalar field is a terrain, where the domain is plane and the function, also called a *height function*, represents elevation, or depth, with respect to the sea level.

Formally, a *d-dimensional scalar field* is defined by a pair (D, f), where D is a d-dimensional domain in $\mathbb{R}^n$, and f is a real-valued function $f : D \to \mathbb{R}$. Commonly, the domain D is connected, and is a d-manifold. Often, D is a d-dimensional hyper-rectangle.

The scalar field represents a physical phenomenon which is measured at a finite set of points in D. Thus, function f is unknown, with the exception of the sampled points. The real shape of D is also unknown, and it is usually approximated by the convex hull, or by the axis-parallel bounding box, of the given point set.

Representing a scalar field in a computational setting implies a discrete approximation of the domain D as well as a discrete approximation of the range of f. These discretizations are built based on the sampled points, in such a way that, given their known values, we are able to provide an estimate of f over the entire (approximation of the) domain D.

The domain is generally partitioned through a d-dimensional simplicial complex, or through a d-dimensional regular grid. Function f is approximated locally within each cell, through some simple analytic function. Thus, a model for a scalar field is called *simplicial* if the domain decomposition is a simplicial complex, while it is called *regular* if the domain is discretized through a regular grid. The cells of a regular grid are known as *pixels* in 2D and *voxels* in 3D, when the scalar field describes a two-dimensional or three-dimensional grey-scale image [13].

Regular models are used when the function values are measured at regularly spaced points. This happens, for instance, with aerial or satellite measurements of terrains or digital 3D scanning of volumes. In a *regular model*, the regular grid is constructed in such a way that measured points (with known field values) are located at the centroids of the d-cells, or at the vertices (0-cells) of the grid. The field values at any other location are interpolated. If the known field values are associated with the d-cells, then a (discontinuous) *step* function is used as field approximation. In this case, we have a so-called *stepped model*. An element (pixel or voxel) within a grid may be considered as connected to the $2d$ neighboring elements lying in the directions of the Cartesian axes (known as *4-connectivity* model for 2D grids), or to the $3^d - 1$ elements lying in the axis-parallel and diagonal directions (*8-connectivity* model for 2D grids) [25]. Figure 1.5 illustrates the two types of grid connectivity.

Fig. 1.5 A pixel (in *black*) and its 4-connected or 8-connected pixels (*shaded*)

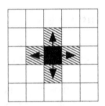 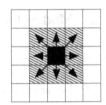

If the known field values are associated with the grid vertices, usually an at least C^0-function is used on the d-cells of the grid. For instance, in 2D we can use a bilinear interpolant over each square cell, or we can divide the cell into two triangles and use linear interpolation over each of them. Many interpolating or approximating functions have been proposed (see, for example, [3, 14]).

Simplicial models [7] are able to deal with irregularly sampled data, such as those obtained from scanning the bounding surface of objects in 3D. Simplicial models are usually constructed with known field values located at their vertices. Within each higher-dimensional simplex, the function is estimated based on a piecewise-linear interpolation of its vertices. Figure 1.6 shows various types of 2D field models.

Regular grids can be encoded in compact data structures, like a matrix of field values. On the other hand, simplicial models require data structures which maintain connectivity information, the relation between d-simplices and vertices, plus adjacency relations among d-simplices [5]. Simplicial models, however, better adapt to variation of the shape, since they can adaptively be built from irregularly distributed data points.

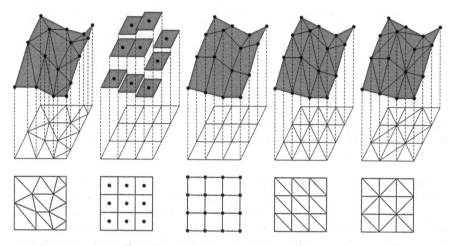

Fig. 1.6 *From left to right*: triangle mesh, regular grid with step function, regular grid with bilinear interpolants, and two differently triangulated grids

1.3 Morse Theory and Morse Complexes

Here, we introduce Morse theory in the smooth case, its relation with morphology, Morse and Morse-Smale complexes, and related properties.

Morse theory captures the relationships between the topology of a manifold, and (the critical points of) a function defined on it. We review here the basic notions of Morse theory in the case of d-manifolds. For more details, see [17, 19].

Let f be a C^2-differentiable real-valued function defined over a manifold domain $M \subseteq \mathbb{R}^d$. A point $p \in M$ is a *critical point* of f if the gradient of f vanishes at p, i.e., if $\nabla f(p) = 0$. Intuitively, this means that the tangent hyper-plane to the graph of f in $\mathbb{R}^{d+1}$ is horizontal. Points which are not critical are called *regular*. Function f is said to be a *Morse function* if all its critical points are non-degenerate, i.e., if the Hessian matrix $Hess_p f$ of the second derivatives of f at p is non-singular (its determinant is $\neq 0$). This implies that the critical points of a Morse function f are isolated.

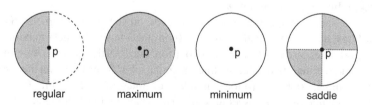

Fig. 1.7 Classification of points in the 2D case. *Shaded regions* indicate points on a small disc around p, with lower function values compared with point p

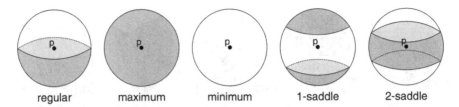

Fig. 1.8 Classification of points in the 3D case. *Shaded regions* indicate points on a small sphere around p, with lower function values compared with point p

The number of negative eigenvalues of $Hess_p f$ is called the *index* of a critical point p. The corresponding eigenvectors show the directions in which f is decreasing. A critical point p is a *minimum* or a *maximum* if it has index 0 or d, respectively. Otherwise, index of p is i, $0 < i < d$, and p is an i-*saddle*. In 2D, there is just one type of saddle, with index equal to 1. In 3D, there are two different types of saddles, corresponding to index equal to 1 and 2, respectively. In Figs. 1.7 and 1.8 we illustrate a neighborhood of critical points in a two and three dimensions, respectively.

The number of different types of critical points has a relation with the Euler characteristic $\chi(M)$ of M:

Theorem 1.1. *Let* $f : M \to \mathbb{R}$ *be a Morse function over a d-manifold* M. *Then*

$$\chi(M) = \sum_{i=0}^{d}(-1)^i c_i,$$

where c_i denotes the number of critical points of f with index i.

An *integral line* of a function f is a maximal path which is everywhere tangent to ∇f. Integral lines follow the gradient directions in which the function has the maximum slope. It can be shown that integral lines are pairwise disjoint, that is, if they share a point, then they are the same line.

Assume to consider integral lines as oriented upwards. Then, no integral line starts from a maximum p, while an arbitrary number of integral lines converge into p. The opposite happens if p is a minimum. If p is a regular point, then there is just one integral line passing through p.

If p is a saddle point, the situation depends on the index of p. In 2D, two integral lines start from p and go to two (non necessarily distinct) maxima, and two integral lines arrive at p from two (non necessarily distinct) minima. In 3D, a 1-saddle s has two integral lines arriving at s from two (not necessarily distinct) minima, and an arbitrary number of integral lines going from s to 2-saddles and maxima. A 2-saddle s has two integral lines going to two (not necessarily distinct) maxima, and an arbitrary number of integral lines arriving at s from 1-saddles and minima.

Integral lines that end at a critical point p of index i, form an i-cell, called a *descending* (or *stable*) *manifold* of p (see Fig. 1.9b). For instance, if M is a

2-manifold, then the descending manifold of a minimum, a saddle, and a maximum, is a 0-cell (vertex), a 1-cell (edge), and a 2-cell (region), respectively.

In a totally symmetric way, integral lines that start from a critical point p of index i form a $(d - i)$-cell, called *ascending* (or *unstable*) *manifold* of p (see Fig. 1.9c). Thus, in 2D, the ascending manifold of a minimum, a saddle, and a maximum, is a 2-cell (region), a 1-cell (edge), and a 0-cell (vertex), respectively.

The ascending (descending) manifolds are open sets, and are pairwise disjoint. They decompose M into open cells which form a complex, since the boundary of every cell is a union of lower-dimensional cells. Such complexes are called *ascending* and *descending Morse complexes*. They are illustrated in Fig. 1.11a and b.

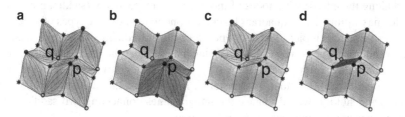

Fig. 1.9 (**a**) Integral lines and critical points of a scalar field f defined on a 2D domain. *Black circles*, *stars*, and *white circles* denote maxima, saddles, and minima, respectively. (**b**) Descending manifold formed by integral lines converging to maximum p. (**c**) Ascending manifold formed by integral lines originating at minimum q. (**d**) Morse-Smale complex with the 2-cell formed by integral lines originating at q and converging to p

A Morse function f is called a *Morse-Smale function* if and only if each non-empty intersection of an ascending and a descending cell is transversal. This means that the intersection (if it exists) of the descending i-dimensional cell of a critical point p of index i, and the ascending $(d - j)$-dimensional cell of a critical point q of index j, $j \leq i$, is an $(i - j)$-dimensional cell. Cells that are obtained as the intersection of descending and ascending manifolds of a Morse-Smale function f decompose M into a *Morse-Smale complex*, illustrated in Fig. 1.9d.

An integral line, which connects two critical points of consecutive index of a Morse-Smale function f, is called a *separatrix*. If we use separatrix lines to cut the domain M of a 2D function f, we can obtain a decomposition of M corresponding to the Morse-Smale complex defined above. For a Morse-Smale function, there is no integral line that connects two critical points of the same index.

In 2D, each 2-cell of a Morse-Smale complex is related to a maximum p and a minimum q, and it is obtained as (a connected component of) the intersection of the descending 2-cell of p and the ascending 2-cell of q. In [9], it has been shown that each such 2-cell is quadrangular, with vertices of index 0,1,2,1 (q, s_1, p, s_2), in this order along the boundary. Saddles s_1 and s_2 are not necessarily distinct, thus it is possible that $s_1 = s_2$. In [20], it has been shown that, for a Morse-Smale function f, there are three different types of 2-cells in the Morse-Smale complex of f, illustrated in Fig. 1.10a–c.

The 1-skeleton of the Morse-Smale complex consists of the critical points and the separatrix lines connecting them. In 2D, it is called a *critical net*.

The *Critical Point Configuration Graph (CPCG)* [20] is a generalization and abstraction of the critical net for a Morse function f defined on the closure of a simply-connected open set in the plane. It is a graph, in which the nodes correspond to the critical points of f and two nodes are connected by an arc if there exists an integral line that emanates from one corresponding critical point and reaches the other. It is not assumed that f is a Morse-Smale function, so some integral lines may connect two saddles. Minimal cycles of edges of the CPCG partition the domain of f into regions, called *slope districts*. There are four different types of slope districts, as illustrated in Fig. 1.10. The first three are quadrangles (possibly glued along the edges) with nodes of index 1,0,1,2 respectively (saddle, minimum, saddle, maximum). These quadrangles correspond to the possible types of 2-cells in the Morse-Smale complex. The first type occurs most frequently in the real data, the second and third type correspond to an isolated mountain (maximun) and an isolated crater (minimum), respectively. The last type of a slope district occurs only if f is not a Morse-Smale function. It is unstable, in the sense that a small perturbation of the scalar field f would replace the integral lines connecting two saddles with integral lines connecting those saddles to extrema.

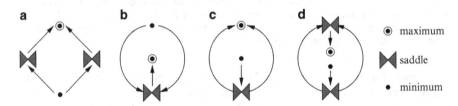

Fig. 1.10 ((a)–(c)) The three possible configurations of 2-cells in a 2D Morse-Smale complex. ((a)–(d)) The slope districts of the CPCG [20]

The *Morse Incidence Graph (MIG)* proposed in [4] is a dual representation for the ascending and the descending Morse complexes Γ_a and Γ_d of a Morse-Smale function. The topology of both complexes is represented by encoding the immediate boundary and co-boundary relations of the cells in the two complexes. The Morse incidence graph provides also a combinatorial representation of the 1-skeleton of a Morse-Smale complex.

A *Morse Incidence Graph (MIG)* is a labeled graph $G = (N, A, \varphi)$ in which:

- the set of nodes N is partitioned into $n + 1$ subsets $N_0, N_1, \ldots, N_n$, such that there is a one-to-one correspondence between the nodes in N_i (*i-nodes*) and the i-cells of Γ_d, (and thus the $(n - i)$-cells of Γ_a);
- there is an arc joining an i-node p with an $(i + 1)$-node q if and only if the corresponding cells p and q differ in dimension by one, and p is on the boundary of q in Γ_d, (and thus q is on the boundary of p in Γ_a) (see Fig. 1.11);

- each arc a connecting an i-node p to an $(i + 1)$-node q is labeled by the number $\varphi(a)$ of times i-cell p (corresponding to i-node p) in Γ_d is incident to $(i+1)$-cell q (corresponding to $(i + 1)$-node q) in Γ_d.

Fig. 1.11 (**a**) Ascending Morse complex, (**b**) descending Morse complex and (**c**) the corresponding Morse Incidence Graph (MIG)

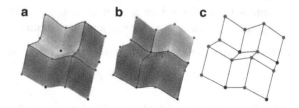

Thus, the MIG captures the combinatorial structure of the critical net, disregarding the geometry of the separatrix lines, as it is a graph in which the nodes correspond to the critical points in the critical net and the arcs correspond to pairs of critical points connected by separatrix lines. In 2D, the (unlabeled) MIG is also known in the literature as the *surface network* [23, 27].

1.4 Watershed Transform in the Smooth Case

We review here the watershed transform, that is an independently developed and alternative framework to Morse theory. The watershed transform has been first defined for grey-scale images and several definitions exist in the discrete case [18, 28]. The watershed transform has also been defined for C^2-differentiable functions over a connected domain D for which the critical points are isolated. This includes Morse functions. Basic notions in the watershed transform are the notions of catchment basin and watershed lines. They can both be defined in terms of topographic distance [18, 24].

Definition 9. If f is a function whose gradient ∇f is non-null everywhere except possibly at some isolated points, then the *topographic distance* $T_D(p, q)$ between two points p, q belonging to the domain D of f is

$$T_D(p, q) = \inf_P \int_P ||\nabla f(P(s))|| ds$$

where P is a path (smooth curve) inside D, such that $P(0) = p$, $P(1) = q$, and $|| \cdot ||$ denotes the magnitude (norm) of a vector.

The topographic distance is defined in this way in order to ensure that the path which minimizes the topographic distance between two points p and q in D is the path of steepest slope, if it exists. In other words, if p and q are two points in D and if there is an integral line which reaches both p and q, then the topographic distance between these two points is equal to the difference in elevation between them (i.e., $T_D(p,q) = |f(p) - f(q)|$). Otherwise, if such an integral line does not exist, the topographic distance between p and q is strictly greater than the difference in elevation between p and q (i.e., $T_D(p,q) > |f(p) - f(q)|$).

Let us consider the set of minima of f. The *catchment basin* $CB(m_i)$ of a minimum m_i is defined as the set of points which are closer (in the sense of topographic distance) to m_i than to any other minimum.

Definition 10. Let m_i be a minimum of f. The *catchment basin* of m_i is

$$CB(m_i) =$$

$$\{p \in D : f(m_i) + T_D(p, m_i) < f(m_j) + T_D(p, m_j), \ m_j \text{ minimum}, \ m_j \neq m_i\}.$$

An alternative definition of catchment basin has been given by Najman et al. in [21].

Watershed (or *watershed lines*) $WS(f)$ of f is defined as the set of points in D which do not belong to any catchment basin, i.e., as the complement in D of the set of catchment basins of the minima of f.

When f is a C^2-differentiable Morse function, then the closure of the catchment basins of the minima of f are the closure of the 2-cells of the ascending Morse complex of f, and the set of watershed lines forms a subset of separatrix lines that connect saddles to maxima. Each catchment basin is bounded by a sequence of saddles, ridge lines and maxima. Symmetrically, if we consider the closure of catchment basins of the opposite function $-f$, then we get the closure of the 2-cells of the descending Morse complex of f.

1.5   Piecewise-Linear Morse Theory

The first attempt to define an equivalent to Morse theory in the discrete case is provided by the piecewise-linear Morse theory, due to Banchoff [1, 2]. Banchoff considers *polyhedral surfaces*, i.e., 2D cell complexes where 1-cells are straight-line segments, 2-cells are polygonal regions, the scalar field is known at the vertices (0-cells) and it is approximated by linear interpolation over other cells. Two-dimensional scalar fields represented through triangle meshes satisfy these requirements. In such a discrete setting, Banchoff defined the theory about critical points, while the notion of a quasi-Morse-Smale complex, introduced by Edelsbrunner [8, 9] captures the characteristics of the Morse-Smale complex in 2D and 3D.

1.5.1   Critical Points in a Piecewise-Linear Model

Let a function f be defined on the vertices of a 2D simplicial complex Σ. Banchoff [2, 22] introduces critical points under the assumption that the values of function f are all distinct. Such a condition has been relaxed in [27] by requiring that function f is *general*, i.e., there are no two adjacent vertices v and w in Σ having the same function value. This means that the graph of f over Σ has no *flat edges*. Under this assumption, critical points may occur only at the vertices of Σ.

The intuitive idea is that a vertex p of Σ can be classified by looking at the configuration of the field f in a small neighborhood around p. Here, a small neighborhood is the fan of triangles incident in p. If the domain of Σ is planar, then function f is a height field, and we can visualize this concept by looking at the horizontal plane through p. The neighborhood of p lies completely below or above such a plane if p is a minimum or a maximum, respectively (see Fig. 1.12b and d). It is split into two pieces if p is a regular vertex (see Fig. 1.12a), and it is split into four or more pieces if p is a simple or multiple saddle, (see Fig. 1.12c). Thanks to the linear interpolation, the number of intersections with the plane is at most one for each triangle incident in p, i.e., at most one for each edge of the link of p (we recall from Sect. 1.1 that the link contains the edges of incident triangles opposite to p). Within the link, edges intersecting the plane are those having one vertex with higher function value than $f(p)$ and the other vertex with lower function value than $f(p)$. Such edges are called *mixed edges*. The number of mixed edges allows to classify vertex p:

- If p has no mixed edges, then it is a *minimum* (if all its adjacent vertices have higher function value than p), or a *maximum* (if all its adjacent vertices have lower function value).
- If p has two mixed edges, then it is *regular*.
- Otherwise, p is a *saddle*. In this last case, the number of mixed edges is even and equal to $2 + 2k$ ($k \geq 1$). Point p is said to be a saddle with *multiplicity k* (or k-fold saddle).

Figure 1.12 shows a regular vertex (a), a minimum (b), a 1-fold saddle (c) and a maximum (d), respectively. A k-fold saddle is called a *simple saddle* if $k = 1$ and *multiple saddle* otherwise. A twofold saddle is also called a *monkey saddle*.

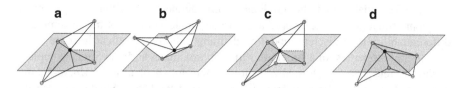

Fig. 1.12 Situation of the link of a vertex with respect to the horizontal plane: (**a**) a regular vertex, (**b**) a minimum, (**c**) a saddle, (**d**) a maximum

The number of intersections with the horizontal plane, that is the number of mixed edges in the link of p, is used to associate a *discrete index* with p, as follows [1, 2]:

$$i(p) = 1 - \frac{1}{2}(\#\{\text{mixed edges for } p\}). \tag{1.3}$$

Discrete critical points are defined as points with index different from 0. In particular, the index is equal to 1 for maxima and minima, while it can assume an arbitrary negative integer value for saddles. For example, a simple saddle will have index equal to -1, a *monkey-saddle* will have index equal to -2, and a k-fold saddle will have index equal to $-k$. The discrete index is different from the classical index (presented in Sect. 1.3), since a Morse function in the smooth case can have only simple saddles, but Banchoff proved the following theorem [1, 2], that holds for general elevation functions defined on polyhedral surfaces.

Theorem 1.2. *Let f be a 2D piecewise-linear function on Σ, such that $f(v) \neq f(w)$ for each pair of mutually adjacent vertices v and w (i.e., f is general). Then:*

$$\sum_{v \text{ vertex} \in \Sigma} i(v) = \chi(\Sigma).$$

The only non-zero contributions to the summation are given by critical vertices, as regular vertices have $i(v) = 0$. Theorem 1.2 also includes the case of isolated degenerate critical points, i.e., multiple saddles, that are not considered by Morse theory. In the discrete case, if f is an elevation function on Σ without multiple saddles, then Theorem 1.2 is equivalent to Theorem 1.1 in two dimensions, and states that the number of extrema minus the number of saddles is equal to $\chi(\Sigma)$. If f has multiple saddles, then each k-fold saddle can be unfolded into k simple saddles (see Sect. 3.1), and the two theorems are again equivalent in the 2D case.

Banchoff also proved the validity of the previous results for general functions defined over d-dimensional complexes, where the *indicator* function generalizes the index defined in Eq. (1.3) [1].

1.5.2 Quasi-Morse-Smale Complexes

In the discrete case, it is not possible to define integral lines because the function is not differentiable and the gradient is not defined. Algorithms to compute (an approximation of) the Morse and Morse-Smale complexes try to simulate differentiability on a discrete model by defining a discrete gradient in some way. The desired properties for a structure, which approximates a Morse-Smale complex in the discrete case, are formalized through the notion of a *quasi-Morse-Smale* complex. It describes the combinatorial structure of the Morse Smale complex in the smooth case, but its arcs and quadrangles (1- and 2-cells) may not be those of maximal ascent and descent. The notion of a *quasi-Morse-Smale complex* in 2D and 3D has been introduced by Edelsbrunner et al. in [8, 9].

In 2D, a *quasi-Morse-Smale complex* is a two-dimensional cell complex Γ, in which the set of vertices (0-cells) can be partitioned into three sets V_{min}, V_{sad} and V_{max} (corresponding to minima, saddles and maxima, respectively, of a

Morse-Smale function f) and the set of edges (1-cells) can be partitioned into two sets A and B, such that:

1. edges in A have one endpoint in V_{min} and one in V_{sad}; edges in B have one endpoint in V_{sad} and one in V_{max};
2. each vertex $p \in V_{sad}$ belongs to four edges, which alternate between A and B in a cyclic order around p;
3. all 2-cells of Γ are quadrangles, with vertices from V_{min}, V_{sad}, V_{max}, V_{sad}, in this order, along their boundary.

Fig. 1.13 (**a**) Combinatorial structure of a 2D Quasi-Morse-Smale complex and (**b**) the corresponding decomposition with colored quadrangles

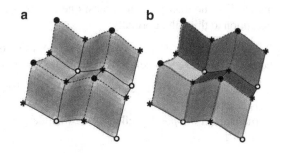

In Fig. 1.13 an example of the combinatorial structure of a quasi-Morse-Smale complex is shown. Vertices of the three sets V_{min}, V_{sad}, and V_{max} are depicted as white dots, stars, and black dots, respectively. Bold lines represent edges in set A while dotted lines represent edges in set B.

The approach to capture the combinatorial structure of a Morse-Smale complex in 3D has been proposed in [8]. In 3D, a *quasi-Morse-Smale complex* is a three-dimensional cell complex Γ, in which the set of vertices can be partitioned into four sets V_{min}, V_1, V_2 and V_{max} (corresponding to minima, 1-saddles, 2-saddles and maxima of a Morse-Smale function f), the set of edges (1-cells) can be partitioned into three sets R, S, and T and the set of 2-cells can be partitioned into two sets P and Q, such that:

1. the edges from R, S, and T connect vertices from V_{min} and V_1, V_1 and V_2, and V_2 and V_{max}, respectively;
2. 2-cells from P and Q are quadrangles with the nodes from V_{min}, V_1, V_2, V_1, and V_1, V_2, V_{max}, V_2, in that order, respectively, around the boundary;
3. each 1-cell in S is in the boundary of four quadrangles, which in a cyclic order alternate between P and Q.

1.6 Forman Theory

Forman theory [11] is a discrete counterpart of Morse theory, and its main purpose is to transpose the results of Morse theory from a smooth to a combinatorial setting. Forman theory can be introduced in terms of a *discrete Morse function* (also called a *Forman function*) defined on all cells of a cell complex Γ, or, equivalently, it can be given in terms of a (negative) discrete gradient vector field (called *Forman gradient vector field*) V defined on Γ [11]. We present here both approaches for cell complexes in which each p-cell γ in the boundary of a $(p + 1)$-cell γ' appears exactly once in the boundary of γ'. Regular grids and simplicial complexes satisfy such a property.

In the first approach, a discrete function F, defined on all the cells (and not only on the vertices) of a cell complex Γ is considered. Such a function F is called a *Forman function* if for any p-cell γ, all the $(p - 1)$-cells in the boundary of γ have a lower F value than γ, and all the $(p + 1)$-cells in the co-boundary of γ have a higher F value than γ, with at most one exception. A cell is *critical* if there is no exception to this rule. Formally:

Definition 11. Let Γ be a cell complex. A function $F : \Gamma \to \mathbb{R}$ is a *Forman function* if, for every p-cell γ, both the following conditions are satisfied:

- the number of $(p + 1)$-cells τ in the co-boundary of γ, such that $F(\tau) \leq F(\gamma)$, is at most 1,
- the number of $(p - 1)$-cells ν in the boundary of γ, such that $F(\nu) \geq F(\gamma)$, is at most 1.

The above two numbers are not both equal to 1.

We observe that, unlike the smooth case, if F is a Forman function on Γ, then $-F$ is not necessarily a Forman function on Γ. Intuitively, critical cells of F are cells where both numbers of Definition 11 are zero.

Definition 12. Let F be a Forman function on a d-dimensional cell complex Γ. A p-cell $\gamma \in \Gamma$ is a *critical cell* of index p if both the following conditions are satisfied:

- for all $(p + 1)$-cells τ in the co-boundary of γ, $F(\tau) > F(\gamma)$,
- for all $(p - 1)$-cells ν in the boundary of γ, $F(\nu) < F(\gamma)$.

If γ is a critical cell, then the index of γ is the same as its dimension. In particular, minima arise at vertices, and if the domain of Γ is a d-dimensional manifold without boundary, then maxima arise at d-dimensional cells [11].

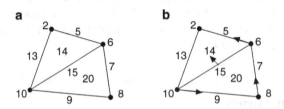

Fig. 1.14 (**a**) A Forman function F defined on a triangle mesh Σ. Each simplex σ is labeled by the value of F at σ. Function F has one minimum vertex (labeled 2), one maximum triangle (labeled 20), one saddle edge (labeled 13). (**b**) The corresponding discrete gradient vector field V_F. An *arrow* $\sigma \to \tau$ is drawn when $(\sigma, \tau) \in V_F$. *Arrows* are of the form vertex-to-edge or edge-to-triangle

Figure 1.14a shows an example of a Forman function F defined on a 2D simplicial complex. Each simplex is labeled by its function value. Vertex labeled 2 is critical (minimum), since function F has higher value on all edges incident to it. Triangle 20 is critical (maximum), since function F has lower value on all edges incident to it. Edge 13 is critical (saddle), since F has higher value on the incident triangle 14, and lower values on its extreme vertices.

Before presenting the second approach to Forman theory, we must introduce the notion of a discrete vector field. Intuitively, a discrete vector field is a collection of arrows, connecting a p-cell of Γ to an incident $(p + 1)$-cell, such that each cell is a head or a tail of at most one arrow. Critical cells are those cells that are neither the head nor the tail of any arrow.

Definition 13. A *discrete vector field* V on a cell complex Γ is a collection of pairs (σ, τ), that we can denote as arrows $\sigma \rightarrow \tau$, such that:

1. σ is a p-cell, and τ is a $(p + 1)$-cell of Γ,
2. σ is a face of τ, and
3. each cell of Γ is in at most one pair of V.

Definition 14. A *V-path* is a sequence $\sigma_1, \tau_1, \sigma_2, \tau_2, \ldots, \sigma_r, \tau_r$ of p-cells σ_i and $(p + 1)$-cells τ_j, $i, j = 1, .., r$, $r \geq 1$, such that $(\sigma_i, \tau_i) \in V$, σ_{i+1} is a face of τ_i, and $\sigma_i \neq \sigma_{i+1}$.

A V-path with $r > 1$ is *closed* if σ_1 is a face of τ_r different from σ_{r-1}.

The arrows in Fig. 1.14b, i.e., $v_6 \rightarrow e_5, v_8 \rightarrow e_7, v_{10} \rightarrow e_9, e_{15} \rightarrow t_{14}$, are elements of a discrete vector field. Here, examples of V-paths are $v_{10}, e_9, v_8, e_7, v_6, e_5$ and e_{15}, t_{14}.

Definition 15. A discrete vector field V is called a *discrete (Forman) gradient vector field* if and only if there are no closed V-paths in V.

Definition 16. A *critical cell* of V of index p is a p-cell γ which does not appear in any pair of V.

Note that the index of a critical cell is always equal to the dimension of the cell.

There is a correspondence between Forman functions and Forman gradient vector fields [10]. Namely, for each Forman function F, a Forman gradient vector field V_F can be constructed. Conversely, for each Forman gradient vector field V there exists a (non-unique) Forman function F such that the gradient field of F is V. Let us explain this correspondence in more detail.

A Forman gradient vector field V_F is obtained from a Forman function F by noticing that non-critical cells come in pairs, and by drawing an arrow from a p-cell γ to a $(p + 1)$-cell τ (adding a pair (γ, τ) to V_F) if γ is a face of τ and $F(\tau) \leq F(\gamma)$. If F is a Forman function, then each cell of Γ is a head or a tail of at most one arrow, and critical cells are those cells that are neither the head nor the tail of any arrow. Thus, for any Forman function F, a corresponding discrete vector field V_F can be constructed, such that along each V_F-path $\sigma_1, \tau_1, \sigma_2, \tau_2, \ldots, \sigma_r, \tau_r$, function F is decreasing, i.e., $F(\sigma_1) \geq F(\tau_1) > F(\sigma_2) \geq F(\tau_2) > \ldots > F(\sigma_r) \geq F(\tau_r)$. This implies that for such discrete vector field V_F, there can be no closed V_F-paths in Γ, i.e., that V_F is a Forman gradient vector field. As in the smooth case, a (negative) Forman gradient vector field V_F of F at a p-cell γ indicates the direction of a unique $(p+1)$-dimensional co-face τ of γ, in which F is decreasing.

In the example in Fig. 1.14b, a Forman gradient vector field V_F of the Forman function F in Fig. 1.14a is illustrated. The pairs of the gradient V_F are $(6,5),(8,7),(10,9),(15,14)$. The critical elements are vertex 2, edge 13, and triangle 20, since there is no arrow starting or ending at them.

Conversely, if a Forman gradient vector field V on Γ is given, then there is a Forman function F such that the gradient field V_F of F coincides with V [11]. Function F can be defined on each p-cell γ of a cell complex through the iterative application of the following rules:

1. if γ is critical, then $F(\gamma) = p$,
2. if a pair $(\gamma, \tau) \in V$ exists, i.e., $\gamma \to \tau$, then $F(\gamma) = p + \frac{d(\gamma)}{2D}$, where $d(\gamma)$ is the length of the longest V-path starting at γ, and D is the maximum length of all V-paths.
3. if a pair $(v, \gamma) \in V$ exists, i.e., $v \to \gamma$, then $F(\gamma) = F(v)$,

In Fig. 1.15a, a Forman gradient vector field V is illustrated, and the length $d(\gamma)$ for each cell γ is indicated. In Fig. 1.15b, a Forman function F corresponding to V is given. Here, $D = 2$.

Fig. 1.15 (a) A Forman gradient vector field V. Numbers indicate the length $d(\sigma)$ of the longest path starting at σ, for each p-simplex σ, $p = 0, 1$. (b) Forman function F, such that the gradient field V_F of F is equal to V

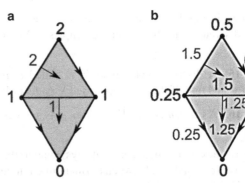

In the combinatorial setup of Forman theory, maximal V-paths correspond to the separatrix lines of a Morse function f in the smooth case, directed downwards. We call *separatrix V-path* connecting critical $(i+1)$-cell τ with critical i-cell σ any V-path $\sigma_1, \tau_1, \sigma_2, \tau_2, \ldots, \sigma_r, \tau_r$ such that σ_1 is a face of τ and σ is a face of τ_r. V-paths mentioned before on Fig. 1.14 are separatrix V-paths.

In [11], the combinatorial counterpart of Theorem 1.1 has been demonstrated.

Theorem 1.3. *Given a Forman function F defined on a cell complex Γ, then*

$$\chi(\Gamma) = \sum_{i=0}^{d}(-1)^i c_i$$

where c_i denotes the number of critical cells of dimension i, also called the Morse numbers *of F.*

The definitions introduced in Forman theory are a discrete analogue of the definitions for smooth functions reviewed in Sect. 1.3. Critical points of smooth Morse functions correspond to critical cells of Forman functions, involved in a number of incoming and out-coming V-paths based on their index.

1.7 Summary

The main notion introduced in this chapter is the *cell complex*. Cell complexes include regular grids and simplicial complexes used to model scalar fields in the discrete case, as well as *Morse* and Morse-Smale complexes used to represent the morphology of scalar fields and their domain in the smooth case. Moreover, we have introduced the *watershed transform*, which is an alternative approach to Morse theory in the smooth case.

We have also discussed different discrete approaches, which transpose the concepts of Morse theory from a smooth to a discrete setting, i.e., the *piecewise-linear Morse theory* by Banchoff and the *discrete Morse theory* by Forman.

These different perspectives, from the watershed transform, to the piecewise-linear Morse theory, and to the discrete Morse theory, lead to different algorithms for morphology computation, as discussed in the next chapters. Table 1.1 provides an overview of the concepts presented in the various sections of this chapter.

Table 1.1 Summary of the concepts revised in Chap. 1

1.1 Some preliminary definitions	Manifold, cell complex, k-skeleton, regular grid, simplicial complex, primal and dual complexes
1.2 Models for scalar fields	Scalar field, regular model and simplicial model
1.3 Morse theory and Morse complexes	Morse function, critical point, integral line, ascending and descending Morse complexes, Morse-Smale complex, Morse Incidence Graph (MIG)
1.4 Watershed transform in the smooth case	Topographic distance, catchment basin
1.5 Piecewise-linear Morse theory	Critical points for piecewise-linear functions, quasi-Morse-Smale complexes
1.6 Forman theory	Forman function, Forman gradient vector field

References

1. T. Banchoff. Critical points and curvature for embedded polyhedra. *J. of Differential Geometry*, 1:245–256, 1967.
2. T. Banchoff. Critical points and curvature for embedded polyhedral surfaces. *American Mathematical Monthly*, 77(5):475–485, 1970.
3. P.A. Burrough and R. A. McDonnell. *Spatial Information Systems*. Oxford University Press, New York, 1998.
4. L. Čomić, L. De Floriani, and F. Iuricich. Building morphological representations for 2D and 3D scalar fields. In E. Puppo, A. Brogni, and L. De Floriani, editors, *Eurographics Italian Chapter Conference*, pages 103–110. Eurographics, 2010.
5. L. De Floriani and A. Hui. Data structures for simplicial complexes: An analysis and a comparison. In Mathieu Desbrun and Helmut Pottmann, editors, *Proc. 3rd Eurographics Symposium on Geometry Processing*, volume 255 of *ACM International Conference Proceeding Series*, pages 119–128, Aire-la-Ville, Switzerland, 2005. Eurographics Association.
6. L. De Floriani, E. Puppo, and P. Magillo. Applications of computational geometry to geographic information systems. In J. R. Sack and J. Urrutia, editors, *Handbook of Computational Geometry*, pages 333–388. Elsevier Science, 1999.
7. L. De Floriani, B. Falcidieno, and C. Pienovi. A Delaunay-Based Method for Surface Approximation. *Eurographics Conference Proceedings* 333–350, 1983.
8. H. Edelsbrunner, J. Harer, V. Natarajan, and V. Pascucci. Morse-Smale complexes for piecewise linear 3-manifolds. In *Proc. 19th ACM Symposium on Computational Geometry*, pages 361–370, 2003.
9. H. Edelsbrunner, J. Harer, and A. Zomorodian. Hierarchical Morse complexes for piecewise linear 2-manifolds. In *Proc. 17th ACM Symposium on Computational Geometry*, pages 70–79, 2001.
10. R. Forman. Combinatorial vector fields and dynamical systems. *Mathematische Zeitschrift*, 228:629–681, 1998.
11. R. Forman. Morse Theory for Cell Complexes. *Advances in Mathematics*, 134:90–145, 1998.
12. J. L. Kelley. *General Topology*. Princeton, N. J.: Van Nostrand, 1955.
13. R. Klette and A. Rosenfeld. *Digital Geometry - Geometric Methods for Digital Picture Analysis*. Computer Graphics and Geometric Modeling. Morgan Kaufmann, San Francisco, 2004.
14. H. Mitasova L. Mitas. Spatial interpolation. In *Geographic Information Systems – Principles, Techniques, Management, and Applications*. Wiley, 1999.
15. A. T. Lundell and S. Weingram. *The Topology of CW Complexes*. Van Nostrand Reinhold Company, New York, 1969.
16. W. S. Massey. *A Basic Course in Algebraic Topology*, volume 127. Springer, 1991.
17. Y. Matsumoto. *An Introduction to Morse Theory*, volume 208 of *Translations of Mathematical Monographs*. American Mathematical Society, 2002.
18. F. Meyer. Topographic distance and watershed lines. *Signal Processing*, 38:113–125, 1994.
19. J. Milnor. *Morse Theory*. Princeton University Press, New Jersey, 1963.
20. L. R. Nackman. Two-dimensional critical point configuration graph. *IEEE Transactions on Pattern Analysis and Machine Intelligence*, PAMI-6(4):442–450, 1984.
21. L. Najman and M. Schmitt. Watershed of continuous functions. *Signal Processing*, 38(1):99–112, 1994.
22. X. Ni, M. Garland, and J. C. Hart. Fair Morse functions for extracting the topological structure of a surface mesh. In *International Conference on Computer Graphics and Interactive Techniques ACM SIGGRAPH*, pages 613–622, 2004.
23. J. L. Pfaltz. Surface Networks. *Geographical Analysis*, 8:77–93, 1976.
24. J. Roerdink and A. Meijster. The watershed transform: Definitions, algorithms, and parallelization strategies. *Fundamenta Informaticae*, 41:187–228, 2000.
25. A. Rosenfeld and A. C. Kak. *Digital Picture Processing*. Academic Press, London, 1982.

26. E. H. Spanier. *Algebraic Topology*. Springer-Verlag New York, Inc., 1966.
27. S. Takahashi, T. Ikeda, T. L. Kunii, and M. Ueda. Algorithms for extracting correct critical points and constructing topological graphs from discrete geographic elevation data. In *Computer Graphics Forum*, volume 14, pages 181–192, 1995.
28. L. Vincent and P. Soille. Watershed in digital spaces: An efficient algorithm based on immersion simulation. *IEEE Transactions on Pattern Analysis and Machine Intelligence*, 13(6):583–598, 1991.

Chapter 2
Morphology Computation Algorithms: Generalities

In this chapter, we consider features common to all algorithms for morphology computation. Such algorithms can be classified based on different criteria, that are in general mutually orthogonal (see Sect. 2.1):

1. *dimension of the input scalar field:*
   some methods are dimension-specific (in practice, either for 2D or for 3D scalar fields), while other methods are dimension-independent.
2. *input format:*
   methods may operate on simplicial models or on regular models; some methods also require specific properties for the input scalar field.
3. *output information:*
   some methods compute (a discrete approximation of) the ascending or descending Morse complex; other methods compute (a discrete approximation of) the Morse-Smale complex.
4. *format of the output information:*
   some methods provide a complete combinatorial description of the output complex; many methods simply provide a classification of the 0-cells (vertices), or of the d-cells (triangles or pixels for $d = 2$, tetrahedra or voxels for $d = 3$).
5. *algorithmic approach:*
   different approaches come from different reference theories: Banchoff's piecewise linear Morse theory, watershed transform in the discrete case, Forman's discrete Morse theory.

The basic component of morphology computation algorithms is the identification of the critical points of the field. Section 2.2 is devoted to a treatment of the techniques for computing critical points. Another general issue with morphology computation algorithms is how to deal with the domain boundary and with regions with same elevation (plateaus). We discuss how to deal with such issues in Sects. 2.3 and 2.4, respectively.

© The Author(s) 2014
L. Čomić et al., *Morphological Modeling of Terrains and Volume Data*,
SpringerBriefs in Computer Science, DOI 10.1007/978-1-4939-2149-2_2

For more details on classifications of morphology computation algorithms, and on computation of critical points, see [5, 8].

2.1 Classification of Morphology Computation Algorithms

In the following, we examine the various criteria which can be used for classifying morphology computation algorithms.

2.1.1 Input Dimension, Format and Properties

Algorithms may have been designed for 2D scalar fields, 3D scalar fields, or they may be dimension-independent. At the same time, they may have been designed for scalar fields represented through regular grids or through simplicial models. Some methods are even independent of the input format, in the sense that they can be applied to both formats, with little change.

Algorithms working on *regular grids* may assume that field values are given at the d-cells (pixels for $d = 2$, voxels for $d = 3$), or at the 0-cells (vertices) of the grid. In the latter case, interpolating functions are used within higher-dimensional cells. Algorithms for *simplicial models* (triangle meshes in 2D, tetrahedral meshes in 3D) always assume that field values are given at the vertices, and linear interpolants are used within higher-dimensional simplices.

In several algorithms, the computation focuses on cells of a certain dimension and is based on navigation among them through adjacency relations. If the computation focuses on 0-cells (vertices), this corresponds to considering the *primal graph* of the input model (see Sect. 1.1.4). If the computation is focused on d-cells, this corresponds to considering the *dual graph* of the input model.

Algorithms for stepped regular models consider the dual graph. Here, the nodes are pixels in 2D, and the arcs are defined according to some connectivity model (4- or 8-connectivity in 2D, see Sect. 1.2 and Fig. 1.5). Algorithms for regular models with interpolating functions work on the primal graph.

Algorithms for simplicial models work on either the primal or the dual graph, depending on the approach (for example, boundary-based methods use the primal graph, and region-growing methods use the dual graph, see Sect. 2.1.3).

Some methods for simplicial complexes require that the input scalar field is general, i.e., all pairs of adjacent vertices have distinct function values. Intuitively, this means that flat edges are not allowed.

2.1.2 Output Information and Its Format

Algorithms may produce a (descending or ascending) Morse complex, or a Morse-Smale complex. In both cases, the output is a cell complex, and thus it consists of cells of all dimensions, and of combinatorial relations of adjacency and incidence between pairs of cells. In practice, such output cell complex can be represented in the form of:

- Just the d-cells of the output complex, represented through a classification of the d-cells (triangles or pixels in 2D, tetrahedra or voxels in 3D) of the input scalar field model, where each d-cell is labeled as belonging to a specific d-cell of the Morse or Morse-Smale complex.
- Just the d-cells of the output complex, represented through a classification of the vertices of the input scalar field model, where each vertex is labeled as belonging to a specific d-cell of the Morse or Morse-Smale complex. A more detailed classification may assign each vertex to an i-dimensional cell of the output complex, where $i = 0, \ldots, d$.
- The skeleton (1-skeleton consisting of points and lines in 2D, 1- and 2-skeletons consisting of points, lines and surfaces in 3D) of the Morse or Morse-Smale complex, with their combinatorial relations. Such skeleton(s) define the boundaries of the d-cells of the output complex.
- A complete representation of the Morse or Morse-Smale complex, i.e., its cells of all three dimensions in 2D or four in 3D, and their combinatorial relations. This corresponds to the Morse incidence graph (see Sect. 1.3), plus geometric information about cells.

Most methods that produce Morse complexes are completely symmetric in the ascending and in the descending case: the ascending Morse complex can be computed as the descending Morse complex by considering the field $-f$ having the opposite function values on the same underlying cell complex. Thus, the two Morse complexes are produced in the same format.

Forman-based algorithms produce the ascending and the descending Morse complexes represented in different formats. Descending cells associated with maxima are expressed as collections of d-cells of the primal complex. Ascending cells associated with minima are expressed in terms of d-cells of the dual complex, i.e., they are collections of vertices of the primal complex.

An algorithm has been presented in [7], which constructs the Morse incidence graph (see Sect. 1.3) in 2D and 3D, starting from a classification of triangles and tetrahedra, respectively.

2.1.3 Algorithmic Approach

Based on the approach they use, the various algorithms can be classified as:

- *Boundary-based* methods, which build the lower-dimensional skeletons of the Morse-Smale complex (boundary lines of 2-cells in 2D, and boundary lines and surfaces of 3-cells in 3D).

- *Region-growing* methods, which build the d-cells (called *regions*) of the descending (ascending) Morse complex, by progressively growing each of them, starting from its seed maximum (minimum).
- *Watershed* methods, originally developed for grey-level images, which compute the ascending Morse complex based on the idea of simulating the diffusion of water.
- *Forman-based* methods, which previously define a Forman gradient from the data.

Boundary-based algorithms build the lower-dimensional skeletons of the Morse-Smale complex. The input can be a regular model or a simplicial model of the scalar field. The 1-skeleton is extracted by computing the critical points and then tracing the integral lines (or their approximations) starting from saddle points and converging to minima and maxima. For 3D scalar fields, the 2-skeleton is also built. Boundary-based algorithms can generate the skeleton(s) of the descending (or ascending) Morse complex by tracing only integral lines from saddles to minima (or maxima).

Region-growing algorithms compute an approximation of the descending (ascending) Morse complex by growing the d-cells, called *regions*, corresponding to the maxima (minima), of the scalar field. Such regions are built as collections of d-cells (triangles in 2D, tetrahedra in 3D) of the scalar field model, by classifying triangles, or tetrahedra. An approximation of the Morse-Smale complex can be obtained as the intersection of the descending and ascending Morse complexes. The input is in general a simplicial model of the scalar field. However, some watershed methods, working on regular models, actually operate in a region-growing fashion.

Watershed algorithms typically work on regular grids considered as stepped models. They generate the ascending Morse complex by labeling each d-cell γ (pixel or voxel) with the minimum p, such that γ belongs to the ascending d-cell of p. They can generate the descending Morse complex by considering scalar field $-f$. Watershed methods based on simulated immersion may label some d-cells as belonging to the skeleton of the ascending Morse complex (these are the so-called watershed cells). They are also able to identify the specific element of the skeleton, even if they do not do that. Watershed algorithms can work on simplicial models as well, by considering the primal graph. In this case, they produce a vertex classification. A vertex classification is produced also by the 3D region-growing algorithm in [15].

Forman-based algorithms construct (either directly, or by first defining a Forman function F) a Forman gradient vector field V and its critical cells starting from a scalar field f. The Forman-based algorithm in [14] can also be classified as boundary-based, as it approximates the 1-skeleton of the Morse-Smale complex by gradient lines connecting the critical cells of F.

Boundary-based and region-growing algorithms are reviewed in Chap. 3. *Watershed* and *Forman-based* approaches are reviewed in Chaps. 4 and 5, respectively.

2.2 Detection of Critical Points

Most algorithms perform the identification of critical points in the input scalar field as a pre-processing step. Some algorithms, specifically boundary-based ones, find critical points of all types (maxima, minima, saddles). Not all algorithms for 3D morphology computation distinguish between 1- and 2-saddles (e.g., [13]), and not all algorithms recognize multiple saddles composed of 1- and 2-saddles (e.g., [6, 28]).

Other approaches find critical points of just one type. For instance, region-growing algorithms find just minima when computing the ascending Morse complex, and maxima when computing the descending Morse complex. They do not explicitly extract saddle points.

Some algorithms do not compute critical points in advance, but they recognize them during the computation. This happens, for instance, in the watershed approach by simulated immersion.

In methods which do not detect saddle points, saddles can be found as a post-processing step, by computing the overlay of the descending and ascending complexes. Let us consider the 2D case. If we have a Morse-Smale function, the 1-cells of the descending complex must intersect transversally the 1-cells of the ascending complex, and the saddles are the intersection points. With non-Morse-Smale functions, the intersection of the 1-cells of the descending complex with those of the ascending complex can include an edge or a chain of edges, and, thus, it is not feasible to detect saddles exactly in this way.

2.2.1 Detecting Critical Points in a Simplicial Model

In this section, we analyze how critical points can be computed on a 2D or 3D simplicial model. Since a simplicial model uses linear interpolants, critical points can only occur at the vertices of the underlying triangle or tetrahedral mesh Σ.

First of all, we define the upper/lower link, and the upper/lower star of a vertex v within a simplicial complex Σ. We recall that the link and star have been defined in Sect. 1.1.2.

- The *upper star* of v is the subset of the star of v containing those simplices σ such that all vertices of σ different from v have a higher function value than $f(v)$.
- The *upper link* of v is the subset of the link of v containing those simplices σ such that all vertices of σ have a higher function value than $f(v)$.
- The *lower star* and the *lower link* of v are defined in a completely symmetric way.

We denote the upper link and the lower link as $Lk^+(v)$ and $Lk^-(v)$, respectively.

A first, and widely used, way to identify critical points uses the definition of critical points on piecewise-linear functions due to Banchoff [4, 20] and discussed in

Sect. 1.5. Details for the 2D and 3D cases can be found in the works by Edelsbrunner et al. [11] and by Takahashi et al. [28], respectively. The following procedure is used to classify a vertex p:

1. Take the *link* $Lk(p)$ of vertex p, and decompose it into:

   - simplices (vertices and edges in 2D; vertices, edges and triangles in 3D) belonging to the upper link $Lk^+(p)$ of p (see Fig. 2.1a for the 2D case).
   - simplices belonging to the lower link $Lk^-(p)$ of p (see Fig. 2.1b for the 2D case).

   Note that, in the 2D case, other edges, which belong neither to the upper nor to the lower link (see Fig. 2.1c) are the mixed edges defined in Sect. 1.5.

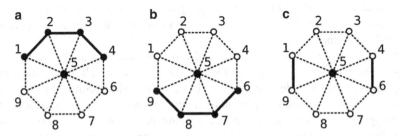

Fig. 2.1 *Bold lines* and *full dots* indicate edges and vertices composing (**a**) the lower link of vertex 5, (**b**) its upper link, and (**c**) the set of its mixed edges of the vertex with field value equal to 5

2. Find the connected components of the upper link of p and of the lower link of p, and count them. Let Δ_+ be the number of components of $Lk^+(p)$, and Δ_- be the number of components of $Lk^-(p)$.
3. Now, classify p:

   - if $\Delta_+ = 0$ (i.e., the upper link is empty), then p is a *maximum*;
   - if $\Delta_- = 0$ (i.e., the lower link is empty), then p is a *minimum*;
   - if $\Delta_+ = \Delta_- = 1$, then p is *regular*;
   - otherwise, p is a *saddle*.

For a 2D complex, in the last case, i.e., when p is a saddle, we always have $\Delta_+ = \Delta_-$, and the multiplicity of the saddle can be computed as $k = (\Delta_+ - 1) = (\Delta_- - 1)$. That is, p is a simple saddle ($k = 1$) if both the upper and the lower link of p have two connected components. It is a multiple saddle otherwise.

For a 3D complex, the type of saddle (i.e., 1-saddle or 2-saddle) can be detected. If $\Delta_+ = 1$ and $\Delta_- = 2$, then p is a simple 1-saddle; if $\Delta_+ = 2$ and $\Delta_- = 1$, then p is a simple 2-saddle. In order to identify multiple saddles, Takahashi et al. [28] count the multiplicity of multiple saddles as follows. If $\Delta_+ = k + 1$, then p is a 1-saddle with multiplicity k. If $\Delta_- = k + 1$, then p is a 2-saddle with multiplicity k.

In 2D, the time complexity of vertex classification, for a triangulation Σ with m vertices, is in $O(m)$. The link of a vertex is a radially sorted sequence of edges and vertices. The connected components of the upper and lower links are found by scanning the radially sorted list and noting the sign changes. Summing up over all m vertices examined, the total number of neighbor pairs is equal to twice the number of edges of Σ, which is in $O(m)$. In 3D, for a tetrahedralization Σ with m vertices, the time complexity is in $O(m^2)$. The link of p is homeomorphic to a spherical surface. Finding the connected components of the upper and lower links requires a graph traversal process, with a linear time complexity in the number of edges of Σ, which can be in $O(m^2)$ in the worst case.

The classical approach to the extraction of critical points, described above, counts the connected components of the upper and lower links of a vertex. However, other approaches have also been proposed.

In Bajaj et al. [3] the classification of critical points in 2D is performed based on vectors normal at the triangles incident in the vertex. Each triangle t has a unit normal vector $\vec{n_t}$ which is normal to the plane of the triangle and points upwards, i.e., $\vec{n_t} = (a_t, b_t, c_t)$ with $c_t > 0$. Let us consider the convex hull of all points (a_t, b_t) corresponding to the triangles in the star of a vertex p. Vertex p is regular or critical depending on whether the convex hull does not contain or contains the origin $(0, 0)$. Thus, according to [3], a point is critical if the normal space of the adjacent triangles includes vector $(0, 0, 1)$.

A more detailed method for 3D scalar fields has been proposed by Edelsbrunner et al. [10]. Their approach is based on (the reduced Betti numbers of) the lower link of a vertex p.

The link $Lk(p)$ of p is a discrete analogue of a sphere around p. Point p is classified based on the reduced Betti numbers $\tilde{\beta}_{-1}$, $\tilde{\beta}_0$, $\tilde{\beta}_1$, and $\tilde{\beta}_2$ of its lower link $Lk^-(p)$. Informally, the Betti numbers β_0, β_1, and β_2 of a simplicial complex Σ indicate the number of connected components, the number of through holes (tunnels), and the number of voids of the domain of Σ. The reduced Betti numbers $\tilde{\beta}_{-1}$, $\tilde{\beta}_0$, $\tilde{\beta}_1$, and $\tilde{\beta}_2$ are the same as Betti numbers, except that $\tilde{\beta}_0 = \beta_0 - 1$ for non-empty complexes, and $\tilde{\beta}_{-1} = 1$ for empty complexes. The same classification of critical points can be obtained by using Betti numbers, but without the symmetry in the classification of minima and maxima.

The classification of a point p is performed as follows:

- if all reduced Betti numbers of the lower link of p are zero, then p is *regular*;
- if $\tilde{\beta}_{-1} = 1$ ($Lk^-(p)$ is empty), and all other reduced Betti numbers are zero, then p is a *minimum*;
- if $\tilde{\beta}_2 = 1$ ($Lk^-(p)$ is equal to $Lk(p)$), and all other reduced Betti numbers are zero, then p is a *maximum*;
- if $\tilde{\beta}_0 = 1$ ($Lk^-(p)$ has two connected components), and all other reduced Betti numbers are zero, then p is a *simple 1-saddle*;
- if $\tilde{\beta}_1 = 1$ ($Lk^-(p)$ is a cylinder), and all other reduced Betti numbers are zero, then p is a *simple 2-saddle*;
- otherwise (i.e., more than one reduced Betti number is different from 0, and/or some of them is larger than 1), then p is a *multiple saddle*.

A multiple saddle p satisfies $\tilde{\beta}_{-1} = \tilde{\beta}_2 = 0$ and $\tilde{\beta}_0 + \tilde{\beta}_1 \geq 2$. Point p is classified as a multiple saddle, composed of $\tilde{\beta}_0$ 1-saddles, and $\tilde{\beta}_1$ 2-saddles. It has been shown that p can be unfolded into $\tilde{\beta}_0$ simple 1-saddles, and $\tilde{\beta}_1$ simple 2-saddles.

2.2.2 Detecting Critical Points in a Regular Grid

For a regular grid, a key issue is the type of interpolation technique used. If the grid is considered as a stepped model, then we can use a characterization of critical points based on counting the connected components of the upper and lower link, as described in Sect. 2.2.1 for simplicial complexes. For example, in the case of a square grid, the field value at a pixel p is compared to the field values of its neighbors, defined based either on the 4- or 8-connectivity model. We have a fixed number of neighbors and, thus, a fixed set of cases. Note that, in case of a 2D grid with 4-connectivity, all saddles will be simple. These approaches [2, 13, 21, 22, 29, 32] are rooted in digital geometry and have been extensively used in image processing [16].

If the regular grid is considered to have field values located at its vertices, then another approach, that we call an *analytic approach*, is used. There is no attempt here at simulating the concept of critical point in the discrete case, but the general idea is that of fitting an interpolating function on the vertices of the grid (at which the field values are known) [2, 24, 25, 30, 31, 33]. The various algorithms differ in the function they use. The selected function usually preserves critical points located at grid vertices, but it may introduce new critical points located inside higher-dimensional cells. Moreover, it is not always globally continuous. Since the interpolating function has a known equation, its critical points can be found analytically. Sometimes they are computed through numerical methods.

The method proposed in [24, 25] uses a bilinear C^0-differentiable interpolating function over a 2D grid, which does not introduce additional minima or maxima, while it may introduce additional saddles inside cells. A grid point p is classified by considering only the elevation of its 4-adjacent neighbors, while a 2-cell, which contains a saddle, is detected by considering the elevation of its four vertices.

In [32], a 3D grid is considered, and a tri-linear interpolating function in each cubic cell is used. In this case also, there may be saddles inside the cubic cells and on their boundaries. The algorithm does not distinguish between 1-saddles and 2-saddles.

Other approaches for 2D or 3D grids [25] use bi-quadratic functions, which provide a globally discontinuous approximation, but guarantee that critical points are constrained to lie at grid vertices. A grid point p is classified based on the characteristics of the functions inside the d-cells incident in p (e.g., in 2D, the function inside a 2-cell can be elliptic, parabolic, or hyperbolic, and a fixed set of cases may occur for p, which are formalized through specific rules).

2.3 Handling the Domain Boundary

A special case in the detection and classification of critical points is represented by the boundary of the domain of the scalar field. Boundary vertices do not have a complete link (homeomorphic to a circle in 2D and to a spherical surface in 3D), but they have an incomplete one (homeomorphic to a segment in 2D and to a disc in 3D). A similar problem arises with boundary pixels and voxels in regular stepped models.

Some methods assume that the domain of the scalar field is a manifold without boundary (e.g., [10]), and, thus, there are no boundary points. However, in most real cases, the domain of a scalar field has a boundary, and thus we must deal with this special case.

The boundary-based methods by Takahashi et al. in 2D [27] and in 3D [28] introduce a virtual minimum (or maximum) which is considered to be adjacent to all boundary vertices of the given simplicial model. This solution, however, causes a non-symmetric treatment of minima and maxima lying on the boundary, depending on the type (minimum or maximum) of the virtual extremum introduced.

Another solution consists of mirroring the function values of adjacent grid points across the boundary (as, for instance, in [13]).

2.4 Presence of Plateaus

Real data often present connected components of vertices (or d-cells in a stepped regular model) all having the same field value. Having in mind the case of a 2D scalar field representing a terrain, these configurations are called *plateaus*, and an edge connecting two vertices with equal field value is called a *flat edge*.

Different solutions are used:

- the notion of a critical point is replaced with that of a critical area, and algorithms use ad-hoc solutions to deal with plateaus (e.g., watershed algorithms consider *regional minima* and other plateaus [18, 19, 23]);
- data are perturbed in a preprocessing step, in order to eliminate flat edges [9, 12].

The first solution implies that connected components of vertices with equal field value (and, thus, of higher-dimensional cells with constant interpolating function) must be identified, classified, and consistently handled. For instance, watershed through simulated immersion [26] can easily identify and treat plateaus during the flooding process. Region-growing methods can identify plateaus which are minima and maxima, and grow regions from them. For plateaus that are not minima or maxima, it is necessary to artificially define one or more entering and/or exiting point [1].

The first solution is very difficult to implement for boundary-based methods, because they should follow lines of steepest slope, which is intrinsically not defined in a plateau. Even in approaches that can be easily adapted to deal with plateaus, the conventions used to process them are somehow arbitrary (especially for plateaus that are not minima or maxima), and may lead to quite different results (as shown in Sect. 7.4.2).

The second solution (i.e., data perturbation) can be achieved by adding random noise to field values, but this introduces many new critical points, which lead to spurious regions in the computed Morse complexes. Thus, a post-processing step is then required to merge such spurious regions.

Another way to perturb data consists of introducing a conventional order among two points with equal function value, in order to decide that one of them is "above" the other one. For instance, a lexicographic order on the spatial coordinates of points can be used [12]. This has the drawback that a regional minimum (or maximum) may give rise to more minima (or maxima), depending on the relative position of points composing it. Again, a post-processing stage is required to merge regions associated with different minima (or maxima) belonging to the same plateau.

Recently, an algorithm for eliminating flat edges from a 2D scalar field in a morphologically consistent way has been proposed by Magillo et al. [17], which represents a valid alternative solution as it does not introduce new critical points. Experiments presented in [17] show that the similarity between the results of different approaches on the same data increases after using such a method as a preprocessing step.

The method is described for terrains (height fields) although it applies to 2D scalar fields in general. It is based on the observation that changing the elevation of a vertex slightly (i.e., of a smaller amount than the minimum elevation difference between v and its adjacent vertices) is sufficient to eliminate flat edges incident in v (by giving a slope to them), while it does not change the uphill or downhill orientation of other edges incident in v.

Plateaus are progressively reduced and eventually eliminated by iteratively changing the elevation of a vertex v lying on the boundary of a plateau. A vertex v is a candidate for this task if either all flat edges incident in v are consecutive around v (the elimination of v does not change the topology of the plateau), or has exactly two incident flat edges (the elimination of v from the plateau either removes a hole or splits the plateau into two). These situations are illustrated in Fig. 2.2.

The highest priority is given to moves which do not change the topology of plateaus, and make v a regular vertex. Such moves are sufficient to eliminate plateaus without holes which are maxima, minima, or regular, or act as simple saddles. Other plateaus need the application of moves where v becomes a saddle, or the topology of the plateau changes. The priority scheme gives preference to cases in which v becomes a saddle with low multiplicity and/or the topology of plateau does not change.

A plateau which was a maximum or minimum gives rise to a set of regular vertices plus a single maximum or minimum vertex, plus as many simple saddles as holes in the original plateau. A regular plateau gives rise to a set or regular vertices. In other cases, a plateau gives rise to a set or regular vertices plus a number of saddle vertices, whose total multiplicity depends on the total number of terrain portions around the original plateau having an elevation above/below the plateau itself, in the original configuration.

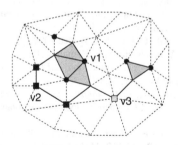

Fig. 2.2 A plateau (formed by *bold edges* and *shaded triangles*) and candidate vertices for elimination. *Black circles* denote vertices, like v_1, which do not change the topology of the plateau. *Black squares* denote vertices, like v_2, which remove a hole from the plateau. *White squares* denote vertices, like v_3, which split the plateau

References

1. L. Arge, J. Chase, P. Halpin, L. Toma, D. Urban, J.S. Vitter, and R. Wickremesinghe. Flow computation on massive grid terrains. *Geoinformatica*, 7(4):283–313, 2003.
2. C. L. Bajaj, V. Pascucci, and D. R. Shikore. Visualization of scalar topology for structural enhancement. In *Proc. IEEE Visualization'98*, pages 51–58. IEEE Computer Society, 1998.
3. C. L. Bajaj and D. R. Shikore. Topology preserving data simplification with error bounds. *Computers and Graphics*, 22(1):3–12, 1998.
4. T. Banchoff. Critical points and curvature for embedded polyhedral surfaces. *American Mathematical Monthly*, 77(5):475–485, 1970.
5. S. Biasotti, L. De Floriani, B. Falcidieno, P. Frosini, D. Giorgi, C. Landi, L. Papaleo, and M. Spagnuolo. Describing shapes by geometrical-topological properties of real functions. *ACM Computing Surveys*, 40(4):Article 12, 2008.
6. Y.-J. Chiang, T. Lenz ans X. Lua, and G. Rote. Simple and optimal output-sensitive construction of contour trees using monotone paths. *Computational Geometry: Theory and Applications*, 30(2):165–195, 2005.
7. L. Čomić, L. De Floriani, and F. Iuricich. Building morphological representations for 2D and 3D scalar fields. In E. Puppo, A. Brogni, and L. De Floriani, editors, *Eurographics Italian Chapter Conference*, pages 103–110. Eurographics, 2010.
8. L. Čomić, L. De Floriani, and L. Papaleo. Morse-Smale decompositions for modeling terrain knowledge. In *Proc. International Conference on Spatial Information Theory (COSIT)*, volume 3693 of *Lecture Notes in Computer Science*, pages 426–444. Springer, 2005.
9. H. Edelsbrunner. *Geometry and Topology for Mesh Generation*. Cambridge University Press, England, 2001.
10. H. Edelsbrunner, J. Harer, V. Natarajan, and V. Pascucci. Morse-Smale complexes for piecewise linear 3-manifolds. In *Proc. 19th ACM Symposium on Computational Geometry*, pages 361–370, 2003.
11. H. Edelsbrunner, J. Harer, and A. Zomorodian. Hierarchical Morse complexes for piecewise linear 2-manifolds. In *Proc. 17th ACM Symposium on Computational Geometry*, pages 70–79, 2001.

12. H. Edelsbrunner and E. P. Mücke. Simulation of simplicity: a technique to cope with degenerate cases in geometric algorithms. *ACM Transactions on Graphics*, 9(1):66–104, 1990.
13. T. Gerstner and R. Pajarola. Topology preserving and controlled topology simplifying multi-resolution isosurface extraction. In *Proc. IEEE Visualization'00*, pages 259–266, 2000.
14. A. Gyulassy, P.-T. Bremer, B. Hamann, and V. Pascucci. A practical approach to Morse-Smale complex computation: Scalability and generality. *IEEE Transactions on Visualization and Computer Graphics*, 14(6):1619–1626, Nov-Dec 2008.
15. A. Gyulassy, V. Natarajan, V. Pascucci, and B. Hamann. Efficient computation of Morse-Smale complexes for three-dimensional scalar functions. *IEEE Transactions on Visualization and Computer Graphics*, 13(6):1440–1447, Nov-Dec 2007.
16. R. Klette and A. Rosenfeld. *Digital Geometry - Geometric Methods for Digital Picture Analysis*. Computer Graphics and Geometric Modeling. Morgan Kaufmann, San Francisco, 2004.
17. P. Magillo, L. De Floriani, and F. Iuricich. Morphologically-aware elimination of flat edges from a tin. In *Proc. 21st ACM SIGSPATIAL International Conference on Advances in Geographic Information Systems (ACM SIGSPATIAL GIS 2013)*, November 5-8 2013.
18. A. Mangan and R. Whitaker. Partitioning 3D surface meshes using watershed segmentation. *Transactions on Visualization and Computer Graphics*, 5(4):308–321, 1999.
19. F. Meyer. Topographic distance and watershed lines. *Signal Processing*, 38:113–125, 1994.
20. X. Ni, M. Garland, and J. C. Hart. Fair Morse functions for extracting the topological structure of a surface mesh. In *International Conference on Computer Graphics and Interactive Techniques ACM SIGGRAPH*, pages 613–622, 2004.
21. L. Papaleo. *Surface Reconstruction: Online Mosaicing and Modelling with Uncertainty*. PhD thesis, University of Genova – Department of Computer Science, 2004.
22. T. K. Peucker and D. H. Douglas. Detection of surface-specific points by local parallel processing of discrete terrain elevation data. *Computer Graphics and Image Processing*, 4:375–387, 1975.
23. J. Roerdink and A. Meijster. The watershed transform: Definitions, algorithms, and parallelization strategies. *Fundamenta Informaticae*, 41:187–228, 2000.
24. B. Schneider. Extraction of hierarchical surface networks from bilinear surface patches. *Geographical Analysis*, 37(2):244–263, 2005.
25. B. Schneider and J. Wood. Construction of metric surface networks from raster-based DEMs. In S. Rana, editor, *Topological Data Structures for Surfaces*, pages 53–70. John Wiley & Sons Ltd, 2004.
26. P. Soille. *Morphological Image Analysis: Principles and Applications*. Springer-Verlag, Berlin and New York, 2004.
27. S. Takahashi, T. Ikeda, T. L. Kunii, and M. Ueda. Algorithms for extracting correct critical points and constructing topological graphs from discrete geographic elevation data. In *Computer Graphics Forum*, volume 14, pages 181–192, 1995.
28. S. Takahashi, Y. Takeshima, and I. Fujishiro. Topological volume skeletonization and its application to transfer function design. *Graphical Models*, 66(1):24–49, 2004.
29. J. Toriwaki and T. Fukumura. Extraction of structural information from gray pictures. *Computer Graphics and Image Processing*, 7:30–51, 1978.
30. L. T. Watson, T. J. Laffey, and R. M. Haralick. Topographic classification of digital image intensity surfaces using generalized splines and the discrete cosine transformation. *Computer Vision, Graphics, and Image Processing*, 29:143–167, 1985.
31. G. Weber and G. Scheuermann. Automating transfer function design based on topology analysis. In G. Brunnett, B. Hamann, H. Müller, and L. Linsen, editors, *Geometric Modeling for Scientific Visualization*, Mathematics and Visualization. Springer Verlag, Heidelberg, 2004.
32. G. H. Weber, G. Scheuermann, H. Hagen, and B. Hamann. Exploring scalar fields using critical isovalues. In *Proc. IEEE Visualization'02*, pages 171–178. IEEE Computer Society, 2002.
33. G. H. Weber, G. Scheuermann, and B. Hamann. Detecting critical regions in scalar fields. In G.-P. Bonneau, S. Hahmann, and C. D. Hansen, editors, *Proc. Data Visualization Symposium*, pages 85–94. ACM Press, New York, 2003.

Chapter 3
Boundary-Based and Region-Growing Algorithms

This chapter describes two approaches to morphology computation which lead to dimension-specific algorithms: the *boundary-based* and the *region-growing* approach. Both approaches have been developed in the context of geographic information systems for 2D scalar fields, intended as height fields representing terrains. In this context, ascending (descending) 2-cells of minima (maxima) correspond to basins (mountains) associated with pits (peaks) of the terrain, and separatrix lines bounding them correspond to ridge (valley) lines. Later, some algorithms were extended to 3D scalar fields, which gained attention in geographic information systems to represent, for instance, atmospheric or geological data, and in scientific data visualization and analysis.

The first approaches developed for height fields used the idea of tracing ridge and valley lines on the terrain. Thus, they used a boundary-based approach. Such algorithms were developed either for regular grids or for simplicial models. Later, region-growing methods were developed for simplicial terrain models, based on the idea of growing a basin or a mountain, starting from a seed minimum (pit) or from a maximum (peak), respectively.

Unlike watershed and Forman-based approaches, which are dimension-independent, both boundary-based and region-growing algorithms are dimension-specific, and they have been designed either for 2D scalar fields or for 3D scalar fields.

In Sects. 3.1 and 3.2, we review boundary-based and region-growing algorithms, respectively, by referring mainly to 2D scalar fields.

© The Author(s) 2014
L. Čomić et al., *Morphological Modeling of Terrains and Volume Data*,
SpringerBriefs in Computer Science, DOI 10.1007/978-1-4939-2149-2_3

3.1 Boundary-Based Algorithms

Boundary-based algorithms compute an approximation of the Morse-Smale complex by computing its 1-skeleton, i.e., an approximation of the separatrix lines connecting critical points of the input scalar field. In 3D, the 2-skeleton, composed of surface patches, is also generated.

All boundary-based methods start by identifying the critical points (minima, saddles and maxima). Then, they extract integral lines by starting from saddles and following (an approximation of) the direction of steepest ascent/descent, until they reach a maximum/minimum.

Boundary-based algorithms exist for simplicial models (triangle meshes in 2D, tetrahedral meshes in 3D) and for regular models (square grids). We review the two cases in the following subsections. Methods reviewed in this section are summarized in Table 3.1.

3.1.1 Boundary-Based Methods on Simplicial Models

The input of these algorithms is a triangle or tetrahedral mesh Σ with manifold domain and field values given at its vertices.

Algorithms use the primal graph of Σ, i.e., the graph whose nodes and arcs correspond to the vertices and edges of Σ, respectively. They extract first the critical points of all types, and then integral lines by starting from saddles. In most methods [2, 12, 21], extracted lines are constrained to follow the edges of the triangle mesh. In [3, 4, 17], the lines are also allowed to traverse the interior of triangles.

In 2D, the output is the 1-skeleton of the Morse-Smale complex. This is a graph where nodes are the critical points (each node is also a vertex of Σ) and arcs are

Table 3.1 Summary of reviewed boundary-based algorithms

Algorithm	Input	Path tracing
Takahashi et al. [21]	2D Simplicial (3D extension in [20])	Based on elevation difference, follow edges
Bajaj and Shikore [2]	2D Simplicial	Based on elevation difference, follow edges
Edelsbrunner et al. [12]	2D Simplicial	Based on slope, follow edges
Bremer et al. [3],Pascucci [17]	2D Simplicial	Based on slope, split triangles
Edelsbrunner et al. [10, 11]	3D Simplicial	Based on elevation difference, follow edges and triangles
Bajaj et al. [1]	2D / 3D Regular	Bi- and tri-cubic interpolation
Schneider [18]	2D Regular	Bi-linear interpolation
Schneider and Wood [19]	2D Regular	Bi-quadratic interpolation

The output of all algorithms is the 1-skeleton of the Morse-Smale complex separatrix lines, which are described as polylines formed by chains of triangle edges (or also by segments crossing the triangles in the output of some methods).

In 3D, the output is the 1-skeleton as above (where separatrix lines are composed of edges of Σ), and the 2-skeleton where each separating surface is a collection of triangles from Σ.

The cell complex, whose skeleton is generated by boundary-based algorithms, may or may not satisfy the conditions of a quasi-Morse-Smale complex. Indeed, not all 2D algorithms guarantee that the regions bounded by the 1-skeleton are quadrangles with alternated saddle-minimum-saddle-maximum on their boundaries, and that all saddles are connected to two minima and two maxima. Moreover, the resulting 2-cells may not be uniformly two-dimensional (they may have dangling edges). Similar problems exist in the 3D case.

3.1.1.1 The Algorithm by Takahashi et al.

The boundary-based algorithm for triangle meshes by Takahashi et al. [21] consists of three main steps:

1. extraction and classification of critical points;
2. unfolding of multiple saddles;
3. tracing of separatrix lines.

In Step 1, the critical points are extracted and classified as maxima, minima and saddles, through Banchoff's method (as described in Sect. 2.2). In each connected component of the upper link of a vertex p, only the vertex with highest elevation is retained. Symmetrically, in each connected component of the lower link of p, only the vertex with lowest elevation is retained. In this way, a reduced list of the neighbors of p is obtained, that will be used for further processing in the next stages of the algorithm. The elements of such a reduced list are called the *representative neighbors* of p.

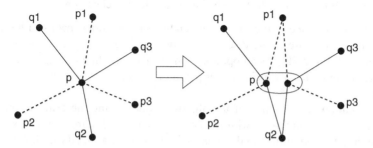

Fig. 3.1 Unfolding a two-fold saddle s, having initially six representative neighbors $p_1, q_1, p_2, q_2, p_3, q_3$, where each p_i has a larger field value than s and each q_i has a smaller field value. A copy of s is created which has four representative neighbors of s (namely q_2, p_3, q_3, p_1) and is a simple saddle. Point s looses two representative neighbors (here p_3, q_3), and its multiplicity decreases to 1 (it becomes a simple saddle)

In Step 2, each multiple saddle is decomposed into a number of simple saddles. This operation is done in order to guarantee that exactly four separatrix lines (two ridges and two valleys) intersect at a saddle point, which is one of the properties of a quasi-Morse-Smale complex.

The procedure to unfold multiple saddles decomposes a k-fold saddle s into k simple saddles. The algorithm iteratively takes four consecutive representative neighbors of s, creates with them a copy of s which is a simple saddle, and deletes two representative neighbors from the original instance of s, thus decreasing its multiplicity. This idea is illustrated in Fig. 3.1.

More precisely, given a k-fold saddle s, the algorithm operates as follows:

- Set $h = k$;
- Let $p_1, q_1, \ldots, p_{h+1}, q_{h+1}$ be the current (circular) list of representative neighbors of s, where points p_i and q_i belong to the upper and lower link of s, respectively;
- while $h > 1$, do the following:

  1. take elements $q_h, p_{h+1}, q_{h+1}, p_1$ from the list;
  2. make a copy of s, mark it as a simple saddle, and set the list of its representative neighbors as $q_h, p_{h+1}, q_{h+1}, p_1$.
  3. eliminate elements p_{h+1}, q_{h+1} from the neighbor list of s;
  4. decrease the value of h to $h - 1$;

Now, all saddles are simple. In Step 3 of the main algorithm, the 1-skeleton of the Morse-Smale complex is constructed. Starting from each (simple) saddle p, four lines are traced in a greedy manner along the edges of the triangle mesh until a minimum or a maximum is reached, each time choosing a highest (lowest) neighbor of the current point. More precisely, for each saddle point p:

1. retrieve the four representative neighbors of p;
2. trace the ridge lines from the two upper representative neighbors up to maxima, choosing at each step the neighbor of highest elevation;
3. trace the valley lines from the two lower representative neighbors down to minima, choosing at each step the neighbor of lowest elevation.

Note that the same portion of a (descending or ascending) path can be traced several times, because in the discrete case two paths going in the same direction may join (see Fig. 3.4b). As an optimization, we can avoiding tracing a path when we reach a vertex from which another path in the same direction has already been traced.

The complexity of Step 1 of the algorithm (identifying the critical points) is in $O(m)$ for a triangle mesh with m vertices. Step 2 (unfolding multiple saddles) requires a time proportional to the total multiplicity over all saddles. This is still in $O(m)$ because the multiplicity of a saddle vertex cannot exceed the number of its incident edges. Step 3 (tracing integral lines) is also in $O(m)$ because the total length of traced (descending or ascending) lines is bounded by the number of edges in the input triangle mesh. This holds under the hypothesis that the algorithm is optimized as explained above.

The algorithm proposed in [2] is similar to the just described one [21]. The difference is that in [2] the classification of critical points is formulated based on vectors normal at the triangles incident in the point (see Sect. 2.2).

3.1.1.2 The Algorithm by Edelsbrunner et al.

The algorithm for triangle meshes by Edelsbrunner et al. in [12] differs from the one in [21] (described in the previous subsection) in the following parts:

- A different criterion is used to define the representative neighbors of a vertex (based on edge slope rather than on difference in field value).
- Unfolding multiple saddles is performed after path tracing. Thus, $k+1$ ascending lines and $k + 1$ descending lines are traced from each k-saddle.
- A final stage is performed in order to improve the slope of traced separatrix lines. This is done through an operation called "handle slide".

Thus, the main steps of the method are:

1. extraction and classification of critical points;
2. tracing of separatrix lines;
3. unfolding multiple saddles;
4. modification of the resulting 1-skeleton through handle slides.

In Step 1, each vertex p of the triangle mesh Σ is classified as regular, minimum, maximum or saddle according to Banchoff's method, as explained in Sect. 2.2. For each connected component of the upper and of the lower link of p, only one representative neighbor is retained, which is defined as the one such that the edge connecting it to p has the maximum slope (whereas [21] retains the one having maximum elevation difference with p).

In Step 2 of the algorithm, starting from each k-fold saddle, $k + 1$ ascending and $k + 1$ descending paths are traced along the edges of the triangle mesh, each time choosing an edge of the steepest ascending or descending slope.

The path tracing algorithms in [12, 21] are similar. The difference is that, in [21], at every vertex, the neighboring vertex with highest elevation is chosen, while in [12], at every vertex, the steepest edge is chosen.

In addition, an elaborate book-keeping procedure is performed to guarantee that two ascending (descending) paths do not split after they have merged, and that there is no intersection of an ascending and a descending path except at a saddle that started them both. It consists of concatenating and duplicating the paths when needed.

The procedure used to unfold multiple saddles in Step 3 has the following difference from the one in [21]. In [21], each step takes a k-fold saddle and decomposes it into a simple saddle and a $(k-1)$-fold saddle, where the latter will be further decomposed if necessary. In [12], each step decomposes a k-fold saddle into two saddles, which have total multiplicity equal to k, and both of them may not be simple. The procedure will iterate on both, if necessary. When duplicating a saddle point s, the algorithm duplicates ascending and descending paths starting from s.

Because the paths are constructed in a greedy manner, by choosing the steepest edge at each vertex, they are not necessarily the ones of steepest ascent or descent. Thus, the result of the previous stages is further modified in Step 4 by using a sequence of local transformations, called *handle slides*. The quadrangular regions are processed in the order of decreasing elevation (where the elevation of a region is the elevation of its lower saddle). A handle slide transforms one quasi-Morse-Smale complex into another, by rerouting the paths within an octagon determined by a quadrangular region $R = abcd$ and two of its adjacent regions, as illustrated in Fig. 3.2.

Fig. 3.2 Handle slide: rerouting of the paths in the octagon $AdaDCbcB$. (**a**) Before and (**b**) after the operation

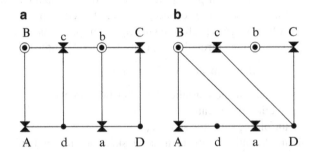

The time complexity of the algorithm is in $O(m)$ for Steps 1, 2 and 3, where m is the number of vertices in the input triangle mesh, according to the same considerations as the algorithm in [21]. The complexity of Step 4 (the number of handle slides) is evaluated in [12] as linear in the number of crossings of separatrix lines between the initial quasi-Morse-Smale complex and the final complex.

The approach in [12] has been extended to computing a quasi-Morse-Smale complex for a three-dimensional scalar field [11]. The resulting algorithm is described in Sect. 3.1.1.4.

3.1.1.3   The Algorithm by Bremer et al.

The algorithm for triangle meshes fields proposed by Bremer et al. in [4] is similar, in its overall structure, to the one in [21]. The difference is that integral lines traced from saddle points do not necessarily follow the edges of the triangle mesh, but they can traverse the interior of triangles. Thus, they are not constrained to the edges of the triangle mesh, but are computed along the actual paths of steepest ascent (descent).

The main steps of the algorithms are:

1. extraction and classification of critical points;
2. unfolding of multiple saddles;
3. tracing of separatrix lines.

The first two steps are performed in a similar way as in [21] (see Sect. 3.1.1.1). We describe now Step 3.

Saddle points are sorted according to the elevation. Two ascending (descending) paths are computed from each saddle, starting from the lowest (highest) saddle p, until they reach a minimum (maximum). The magnitude of the gradient is computed for each edge and for each triangle in the star of p, and the descending (ascending) path from p is continued in the direction where the magnitude of the gradient is maximal. Thus, it may continue either along an existing edge or across a triangle. In the latter case, the triangle is split.

Similar criterion for the continuation of a path is given, when the current path vertex is on the edge of the triangle mesh. In this case, the gradient along the edge, and across the two adjacent triangles, is compared.

If the steepest slope happens to be at an edge, then the edge chosen in [3] is the same as the one chosen in [12]. But, differently from [12], here the path follows the actual direction of steepest ascent or descent, by going across triangles when necessary.

The ascending paths are computed in a similar way, by considering the descending paths of the function $-f$.

The details of the computation of the paths of steepest ascent (descent) are described by Pascucci et al. in [17].

The time complexity of Steps 1 and 2 is in $O(m)$ as in [21]. The complexity of Step 3 is proportional to the total number of segments drawn inside triangles. Note that more segments (all parallel to the unique direction of steepest slope) can be drawn in each triangle t, arriving at t from different points. De Berg et al. [7,8] have shown that the complexity of the river network on a triangulated terrain can be up to $O(m^2)$ in the worst case, but it is in $O(m)$ in real cases. Thus, the overall time complexity is in $O(m)$ in practical cases.

Although more precise, tracing separatrix lines inside triangles increases the computational complexity. De Berg et al. [9] compare various practical methods that trace separatrix lines only along edges, with the exact one tracing separatrix lines inside triangles as well. They also present a hybrid approach which tries to balance computational time and precision.

3.1.1.4 The Algorithm for 3D Scalar Fields by Edelsbrunner et al.

The algorithm proposed by Edelsbrunner et al. in [11] for tetrahedral meshes, computes the 1- and 2-cells which bound the 3-cells in the Morse-Smale complex. The extracted complex has the correct combinatorial structure described by a quasi-Morse-Smale complex. Each 3-cell in the extracted complex has quadrangular faces.

The quasi-Morse-Smale complex is constructed during two sweeps over the input simplicial complex Σ. The first sweep (in the direction of decreasing function value) computes the descending 1- and 2-cells and the second sweep (in the direction of increasing function value) the ascending 1- and 2-cells of the Morse complexes.

During the first sweep, the descending 1-cells and 2-cells are computed simultaneously. Let p be the current vertex in the sweep. If p is a 1-saddle, then the descending 1-cell associated with p is built. Such a 1-cell consists of two chains of edges starting at p and ending at two (not necessarily distinct) minima. It is built as follows:

- The 1-cell of p is initialized by taking the two edges connecting p to the lowest vertex in each connected component of the lower link of p, as illustrated in Fig. 3.3a.
- If a descending arc arrives at a minimum q, then it ends at q.
- If a descending arc arrives at a non-minimum vertex q, then the descending 1-cell is continued by adding the edge from q to the lowest vertex in its lower link (note that, in the discrete case, q is not necessarily regular).

  Indeed, if q is a 1-saddle, some additional operations are done to prepare the ascending 2-cell which will be later built from p (here, details are omitted for brevity).

Note that, in the discrete case, a descending path may traverse other saddle points before ending at a minimum. The used rule ensures that, after two descending paths have joined, they never split.

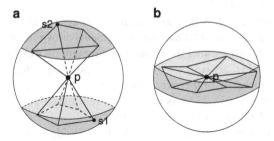

Fig. 3.3 (**a**) The descending 1-cell associated with 1-saddle p is initialized by connecting p to the two lowest vertices s_1 and s_2 in its lower link in [11]. (**b**) The descending 2-cell associated with a 2-saddle p is initialized by the triangles determined by p and a cycle of edges in the lower link of p in [11]

Again, let p be the current vertex p in the sweep. If p is a 2-saddle, then the descending 2-cell of p is built. Such 2-cell consists of a collection of triangles from Σ, which is progressively expanded from its current boundary. Edges of the current boundary are either frozen (i.e., final) or unfrozen (i.e., the 2-cell must expand from them). The descending 2-cell of a 2-saddle p is built as follows:

- To initialize the 2-cell γ of p, a cycle of edges encircling the lower link is constructed, which contains the lowest vertex in the lower link of p. Triangles determined by p and edges of such a cycle form the initial 2-cell γ, as illustrated in Fig. 3.3b. Initially, the entire boundary of γ is unfrozen.
- Cell γ is expanded by constructing a shortest-path tree in the lower link of the current (highest) vertex q on the unfrozen boundary of γ, and connecting q to the edges of this tree. If q is a critical point (a 1-saddle or a minimum), it is declared frozen together with its two incident edges on the boundary.
- When the complete boundary of a 2-cell is frozen the 2-cell is completed.

In the second sweep, ascending 1-cells and 2-cells are constructed piecewise inside the descending 2-cells and 3-cells.

At each 2-saddle p, the construction of an ascending 1-cell is started. An ascending 1-cell is built in a symmetric way with respect to a descending 1-cell in the first sweep, with the difference that traced paths must not cross any already established descending 2-cell.

At each 1-saddle p, an ascending 2-cell is built similarly to descending 2-cells during the first sweep. The difference is that any ascending 2-cell is decomposed into quadrangles by the (pre-computed) intersection curves between descending and ascending 2-cells, and by the ascending 1-cells. Thus, ascending 2-cells are built one quadrangle at a time.

The running time of the algorithm is in $O(m \log m + m_i + m_o)$, where m is the number of vertices of Σ (and the logarithmic factor is due to sorting of vertices by function value), m_i is the input size, i.e., total number of vertices and triangles of Σ, and m_o is the output size (which may exceed the input size, since a simplex in Σ may belong to several descending/ascending cells, and it may belong several times to a single cell).

Edelsbrunner et al. propose another algorithm in [10] for building the descending Morse complex of a 3D scalar field represented through a simplicial model. In some sense, the algorithm is both watershed-based and boundary-based. The vertices are processed in the decreasing order of function values. After labeling the simplices in the lower link of the current vertex that belong to the boundary between descending 3-cells, it marks the simplices that belong to 3-cells. The final output of the algorithm are the 3-cells in the descending Morse complex, which are topological cells, i.e., homeomorphic to a ball.

3.1.2 Analysis and Comparisons

Since the saddles of triangle mesh may be multiple saddles, in general $(k + 1)$ ascending integral lines, and $(k + 1)$ descending integral lines should start from a k-fold saddle. This would lead to a violation of the properties of a quasi-Morse-Smale complex, according to which a saddle must have exactly four incident separatrix lines (namely two ridges and two valleys). For this reason, boundary-based methods unfold multiple saddles to decompose each of them into a number of simple saddles.

Starting from the saddle points, all methods construct separatrix lines as lines of steepest descent and lines of steepest ascent. The methods in [2, 12, 21] compute them along the edges of the triangle mesh, while the methods in [3, 17] let them cross the triangles.

If separatrix lines are constrained to follow the edges of the input triangle mesh, then the result is not the 1-skeleton of a Morse-Smale complex because some extracted lines may connect two saddles (see Fig. 3.4a). This violates the conditions of a quasi-Morse-Smale complex because the sequence saddle-saddle is not allowed on the boundary of a 2-cell. Moreover, two ascending or descending paths may join before reaching the final maximum or minimum (see Fig. 3.4b). This is an intrinsic limitation due to the discretization of the scalar field, as a vertex in the triangle mesh may have fewer edges than the number of separatrix lines which should converge at it. As a result, the constructed 2-cells may not be uniformly two-dimensional and may have dangling one-dimensional parts. In Fig. 3.4b, two descending paths overlap from the vertex at elevation 3 to their final minimum at elevation 0. Finally, the same portion of a path can be traversed both as an ascending and as a descending path, starting from two different (and connected) saddles (see Fig. 3.5a), and an ascending and a descending path may intersect in a point which is not a saddle.

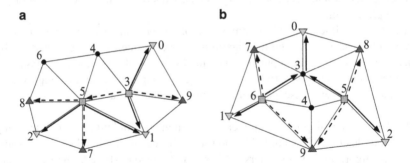

Fig. 3.4 *Green squares* mark saddle vertices, *plain* and *dashed arrows* denote descending and ascending paths, respectively, traced from saddles by the algorithm in [21]. (**a**) The ascending path from saddle at elevation 3 reaches the saddle at elevation 5; the 2-cell of minimum at elevation 1 has two consecutive saddles on its boundary. (**b**) The two descending paths from saddles at elevations 6 and 5 join at a vertex before reaching the minimum at elevation 0; the 2-cell of the maximum at elevation 9 has dangling edge 3–0. (Color figure online)

The algorithm in [21] depends only on the field values and on the connectivity of the triangle mesh, while the actual shape of the triangles does not affect the result. The methods in [3, 12] consider the actual slope of edges (and possibly of triangles), so the geometry underlying triangulation is relevant for the result. Because of this, the two approaches in [21] and [12], both tracing separatrix lines along triangle edges, produce two different critical nets, as illustrated in Fig. 3.6. In [12], the saddle at elevation 2 is connected to the maximum at elevation 20, while in [21] it is connected to the maximum at elevation 21 (if the distance between points at elevation 2 and 20 is sufficiently smaller than the distance between points at elevation 2 and 21, so that the corresponding slope is greater).

The advantage of tracing separatrix lines across the interior of triangles, as done by Bremer et al. [3], is the extraction of the actual steepest lines. The algorithm explicitly forbids merging of two paths going in opposite directions, i.e., the situation in Fig. 3.5a cannot arise. Thanks to this fact, the quadrangular 2-cells obtained have connected interiors, i.e., the situation of Fig. 3.5b cannot occur. However, it is still possible that two paths, which are both ascending or both descending, join, as in Fig. 3.4b.

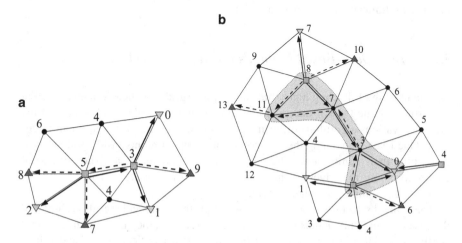

Fig. 3.5 *Green squares* mark saddle vertices, *plain* and *dashed arrows* denote descending and ascending paths, respectively, traced from saddles. (**a**) The edge connecting saddles at elevation 5 and 3 is traversed both as an ascending and as a descending path. (**b**) A Morse-Smale region with disconnected interior. (Color figure online)

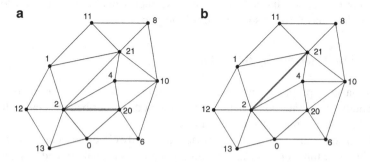

Fig. 3.6 (**a**) The saddle at elevation 2 is connected to the maximum at elevation 20 in [12]. (**b**) It is connected to the maximum at elevation 21 in [21]

Unlike region-growing methods, boundary-based algorithms guarantee that saddles are on the 1-cells of the output complex, but the 2-cells may be non-manifold and not uniformly two-dimensional (see Fig. 3.4b).

Boundary-based algorithms assume that the triangle mesh has no flat edges. If flat edges are present, as it happens with most real data sets, Takahashi et al. [21] suggest to introduce a difference by using a lexicographical order on the xy-coordinates of vertices having equal field value. However, this procedure may introduce extra critical points (it may generate more minima for a single minimum plateau), and thus lead to over-segmentation in the output (see Sect. 2.4). Edelsbrummer et al. and Bremer et al. [3, 12] use the *simulation of simplicity* technique [13], where the idea is an infinitesimal perturbation of data.

3.1.3   Boundary-Based Methods on Regular Grids

The input of the following algorithms is a regular grid with field values located at its vertices. Some type of interpolating function is used within each square of the grid. The various algorithms differ in the function they use. Two important issues exist. One issue is avoiding the creation of new critical points inside grid squares, or keeping their number low. The other issue is guaranteeing a certain degree of continuity along the edges of the grid. These two issues are in conflict and the various algorithms try to balance them.

The general structure of boundary-based algorithms on regular grids is similar to the case of simplicial models (revised in Sect. 3.1.1), i.e., they trace separatrix lines starting at saddles. The difference is that they extract critical points, and follow separatrix lines, by referring to analytic functions defined on the grid.

The output is the 1-skeleton of an approximation of the Morse-Smale complex. Similarly to Sect. 3.1.1, this is a graph where nodes correspond to the critical points and arcs, joining pairs of critical points, correspond to sequences of segments. Depending on the specific algorithm, such segments may be constrained to grid edges, or they may traverse square cells of the grid.

3.1.3.1   The Algorithm by Bajaj et al.

In [1], Bajaj et al. proposed a method for 2D or 3D grids, using bi-cubic interpolation in 2D and tri-cubic interpolation in 3D. The method is described in detail in the 2D case and results are presented in both dimensions.

In the 2D case, the interpolating surface is a globally C^1-differentiable Bernstein-Bézier bi-cubic function. The aim is to interpolate the original gridded data, in such a way that:

- the gradient at the boundaries of the cells is C^0-continuous;
- no critical points are removed, and the number of critical points introduced is kept small.

If the derivatives were computed by classical central differencing, a unique C^1-differentiable bi-cubic interpolant would be obtained, but this interpolant is likely to introduce a large number of (new) critical points. So, the authors develop a "damped" central differencing scheme which they use to compute the first and second order partial derivatives of the interpolating function. The basic idea of "damping" is to keep the interpolant monotone inside grid cells whenever possible.

Integral lines are computed following a Runge-Kutta technique. From each saddle point four integral lines are traced in the direction of the appropriate eigenvectors. Computation of an integral line ends when the line reaches a neighborhood of another critical point or the boundary of the domain. Degenerate critical points and plateaus are ignored.

The algorithm extracts a critical net which is not the 1-skeleton of a Morse-Smale complex, since it uses a C^1-differentiable (and not C^2-differentiable) interpolant. Moreover, the method can generate new critical points with respect to the initial point set.

The algorithm is used in [1] for the enhancement of visualization tools for rendering both 2D and 3D scalar fields.

3.1.3.2 The Two Algorithm by Schneider and Wood

The first method proposed by Schneider and Wood for square grids in [18, 19] uses, for each 2-cell of the grid, a bilinear C^0-differentiable interpolating function of the form $f(x, y) = axy + bx + cy + d$. Coefficients a, b, c, d are obtained unambiguously for each 2-cell from the elevation values of the vertices of the 2-cell. This interpolating function cannot introduce additional minima or maxima (for $a \neq 0$), so minima and maxima can only occur at the vertices of the square grid, but it may introduce additional saddles inside cells at a point with coordinates $(-\frac{c}{a}, -\frac{b}{a})$. A grid point p is classified by considering only the elevation of its 4-adjacent neighbors, while a 2-cell, which contains a saddle, can be detected by considering the elevation of its four vertices.

Separatrix lines are traced and constructed point by point and they can follow grid edges, or go through 2-cells. When a separatrix line crosses a square cell, it can be approximated with small (linear) steps, or computed exactly, by solving a linear system of differential equations. The exact solution (integral line) is a hyperbolic function inside a square.

Unlike the method in [1], this method computes the first and second derivatives analytically. Then, a step-by-step numerical procedure is used to trace separatrix lines.

The second method for square grids described by Schneider and Wood in [19] uses, for each 2-cell of the grid, a bi-quadratic approximation of the form $f(x, y) = ax^2 + by^2 + cxy + dx + ey + f$. This approximating function is constructed by fitting a bi-quadratic polynomial to the 8-connected neighbors to each point of the grid. The method produces a globally discontinuous approximation, formed by local surface patches. In this approach, all the critical points are constrained to lie on the grid vertices.

The coefficients of the bi-quadratic polynomial are determined by least square differences, using a window around each grid point p. In this way, scale dependency is introduced, as a larger window corresponds to a smaller level of scale.

On the basis of the coefficients of the bi-quadratic polynomial, each surface patch is classified as *elliptic, parabolic* or *hyperbolic*, if $4ab - c^2$ is $> 0, = 0$, or < 0, respectively.

The algorithm works in three main steps:

1. Extract the critical points. Each vertex p is classified, based on the type of the corresponding conic section and the number of intersections of the semi-axes with a circle around p with user defined radius r, called the *region of interest* (see [19] for details).
2. Starting at the saddles, trace two paths of steepest ascent and two paths of steepest descent as follows:

   - If no semi-axis intersects the region of interest, follow the gradient direction,
   - If one semi-axis intersects the region of interest, move parallel to the semi-axis,
   - If two semi-axes intersect the region of interest, this means that a minimum or maximum has been reached (or the line hits the border of the grid). Thus, stop tracing.

The extracted critical net depends on the thresholds used to classify the grid points, on the size of the window, and the radius of the region of interest. This may cause topological inconsistencies in the extracted network. Various post-processing heuristics are proposed to correct such inconsistencies. For example, if there is no valley line separating two maxima, then the two maxima can be topologically merged.

As the method described before from [18, 19], this algorithm computes the first and second derivatives analytically. Then, it uses this information to trace the separatrix lines, while the first method in [19] proceeds in a step-by-step numerical manner.

3.2   Region-Growing Algorithms

Region-growing algorithms extract an approximation of the descending (or ascending) Morse complex. In this context, the maximal cells of the descending (ascending) Morse complex are called *regions*. Regions are computed by progressively growing them around the maximum (minimum) they are associated with. Algorithms properly denoted as "region-growing" operate on simplicial models. However, algorithms following a region-growing paradigm, and operating on regular grids, can be found in the watershed approach. For instance, the simulated immersion method (see Sect. 4.1) can be seen as a region-growing paradigm and it can be applied to a regular grid.

Region-growing algorithms work on a simplicial model Σ with a function f whose values are given on the vertices of Σ. They operate on the dual graph

$G = (N, A)$ of the simplicial model, i.e., on the graph whose nodes correspond to d-simplices (triangles in 2D, tetrahedra in 3D), and two nodes are connected by an arc if and only if the two corresponding d-simplices are adjacent along a $(d - 1)$-dimensional face.

Region-growing algorithms for 2D and 3D scalar fields compute the ascending or the descending Morse complex by building its 2-cells as collections of triangles from Σ, or 3-cells as collections of tetrahedra, respectively. The result of the algorithm is a triangle classification in 2D, or tetrahedron classification in 3D, where each triangle or tetrahedron t is labeled with the (minimum or maximum) vertex p, such that t belongs to the (ascending or descending) 2-cell or 3-cell of p.

The only exception is the 3D algorithm by Gyulassy et al. [14]. This algorithm uses the primal graph of the scalar field model and computes the i-cells of the Morse-Smale complex (for all dimensions $i = 0, 1, 2, 3$) by vertex labelling.

We describe here how the algorithms compute the descending Morse complex. The ascending one is computed in a completely symmetric way. In the following, the term *region* is used as a synonym for a d-cell of the descending Morse complex.

3.2.1 The Two Algorithms by Danovaro et al.

Two region-growing algorithms for triangle meshes have been proposed by Danovaro et al. in [5,6]. The general approach is the same. They start a region from a seed that is the highest unprocessed vertex, but is not necessarily a maximum. After the 2-cell of the current seed has been built, it is possibly merged with an already existing and adjacent 2-cell, if the seed was not a maximum. The key point is that seeds are processed in decreasing order of field value.

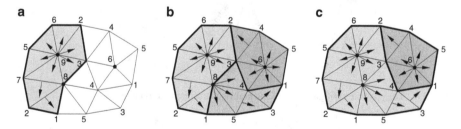

Fig. 3.7 (**a**) Construction of the descending 2-cell γ for vertex at elevation 9. An *arrow* pointing to a triangle from an edge (or from a vertex) means that it is included into γ when reached from that edge (or in the initialization of γ from the seed point). (**b**) Construction of the 2-cells of vertices at elevations 8 and 6. (**c**) The 2-cell of vertex at elevation 8 is merged into the one of vertex at elevation 9, because its seed vertex lies on the boundary (it is not a maximum)

The algorithm in [6] works as follows:

1. Sort all vertices by decreasing elevation.
2. Mark all triangles as unlabeled.

3. Iteratively pick the highest unprocessed vertex p and do the following:

   a. Initialize the tentative descending 2-cell γ of vertex p, with the unlabeled triangles among those in the star of p.
   b. Grow γ by adding an adjacent triangle according to the following criterion: the current 2-cell γ is extended to include the triangle $t = abc$ which is adjacent to γ along the edge ab if the remaining vertex c of t is at lower elevation than a and b.
   c. Stop the region growing process when 2-cell γ cannot be extended any more.

4. For each constructed 2-cell γ, if its seed vertex p lies on the boundary of another 2-cell γ', then merge γ into γ' (p was not a maximum). If p lies on the boundary of more than one 2-cell, choose γ' as the one corresponding to the highest maximum.

The region-growing process is illustrated in Fig. 3.7.

The algorithm in [5] has the same general structure, but it uses a discrete gradient defined on triangles, whereas [6] simply considers the difference in vertex elevation.

The (negative) gradient vector for a triangle t is computed as the gradient of the linear function which interpolates f within t. Each triangle t is traversed by a *bundle of integral lines* which are parallel to the gradient vector. Every edge of a triangle is labeled by the algorithm as being an *entrance* or an *exit* edge (see Fig. 3.8), based on the sign of the inner product of the gradient vector and the outer normal to the edge. If a triangle has two exit edges, then the best exit is determined with a heuristics which considers the size of the angle between the normals to the edges and the gradient vector of the triangle (as illustrated in Fig. 3.9).

Fig. 3.8   The possible configurations of integral lines traversing a triangle

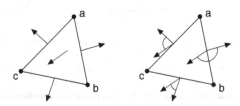

Fig. 3.9   A triangle $t = abc$ with its gradient, and the outer normals to its three edges. Edge ab is an entrance, while ac and bc are exits. The best exit is bc, since its normal minimizes the angle with the gradient of t (i.e., maximizes the inner product)

After this, the algorithm performs the same steps as the algorithm in [6]. The criterion for extending the current 2-cell γ considers an edge e on the boundary of γ. If e is an edge, shared by triangles t and t', and if e is the best exit of triangle t which is inside γ, then the descending 2-cell γ is extended across e to include t' if e is the entrance of t'.

A kind of discrete gradient vector field can be defined starting from the steps performed by the algorithm to include triangles into 2-cells, by taking all pairs (e, t) where triangle t has been added to a 2-cell from edge e, and all pairs (p, t) where p is a seed vertex of a 2-cell and t is an incident triangle added to the 2-cell of p.

If m denotes the number of vertices in the input triangle mesh, the time complexity of both algorithms is in $O(m \log m)$. Such complexity is dominated by the time needed to preliminarly sort the vertices by field value (Step 1). In the subsequent region-growing process (Step 3), each vertex is examined once, each triangle can be considered at most three times (i.e., from each of its three edges) before being added to some 2-cell, and a triangle is added to just one 2-cell. Thus, the time complexity of this stage is in $O(m)$. Step 4 (merging) is linear in the number of the 2-cells (and the number of triangles in such 2-cells) built at Step 3, i.e., it is in $O(m)$.

The algorithm in [6] works in three dimensions with time complexity in $O(m^2)$. It seems not difficult to extend the method to higher-dimensional scalar fields as well.

The description of the algorithms in [5, 6] assumes that no flat edges are present in the input triangle mesh. If they are present, the authors suggest to remove them by perturbing the data in a preprocessing step. However, this may lead to the creation of new minima and maxima and, thus, to over-segmentation (see Sect. 2.4).

On the other hand, it is not difficult to extend the algorithm to deal with flat edges. Indeed, the special case here is represented by plateaus composed of flat triangles. These can be detected in a preprocessing step. When the next highest seed p is picked, and p is on the boundary of a plateau, then the 2-cell of p is initialized with the entire plateau. The subsequent merging of 2-cells will guarantee that the 2-cells constructed from non-maxima plateaus are merged to some adjacent 2-cell.

3.2.2 The Algorithm by Magillo et al.

Another region-growing algorithm for 2D scalar fields has been presented in [15]. Initially, the algorithm labels the vertices of each triangle t based on their relative elevation: the highest, lowest, and middle vertex of t is labeled as S, D, and T, respectively (the intuition behind such labeling is that water flow traversing t has source in S, drain in D, and passes through T).

Then, the following steps are performed:

1. Extract the maxima, that are vertices labeled S in all their incident triangles.
2. As long as there is a maximum vertex p that has not yet been processed, take p as the seed for a new descending 2-cell γ, and build γ in the following way:

   a. Initialize γ with all the triangles incident in p.
   b. Iteratively examine each edge e of the current boundary of γ, and check whether the triangle t, externally adjacent to e, can be added to γ. Criteria used for the inclusion of t are illustrated later.
   c. When no triangle can be further added, stop.

To illustrate the region growing criteria, let $e = ab$ be the current boundary edge of 2-cell γ, and let $t = abc$ be the triangle to be tested for inclusion into γ. Triangle t is not included if vertex c is labeled S in t (see Fig. 3.10a). If c is the labeled D in t, then it is included (see Fig. 3.10b).

If c is labeled T in t, we note that the S-vertex of t is one of the two endpoints of e, say a. The algorithm assumes to have included t and, under this assumption, repeats the inclusion test with edge $e' = ac$ and the triangle t' externally adjacent to t along e'. This process ends when a triangle t_{end} is found such that its third vertex c_{end} is either the D-vertex or the S-vertex of t_{end}. In the first case, all the triangles in the traversed chain from t to t_{end} are included (see Fig. 3.10c). In the second case, no triangle is included (see Fig. 3.10d).

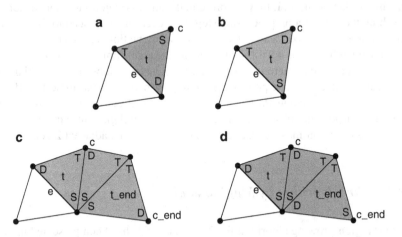

Fig. 3.10 Test used in [15] for the inclusion of a triangle t, sharing an edge e with a triangle already belonging to γ (*white triangle*). In (**a**) triangle t is not included as its third vertex is labeled S in t. In (**b**) t is included as its third vertex is labeled D in t. In (**c**) t is included along with all *green triangles*. In (**d**) no triangle is included. (Color figure online)

The time complexity of the algorithm is in $O(m)$, where m is the number of vertices in the input triangle mesh. Vertex labelling requires a single scan of all triangles, and maxima are found with a single scan of all vertices, in $O(m)$ time.

The cost of Step 3 is also in $O(m)$, because each triangle is tested at most three times, and it is added to exactly one 2-cell.

The description, given above, assumes that no flat edges exist in the input triangle mesh. In [15], rules are provided to handle flat edges and flat triangles as special cases. Plateaus, i.e., connected components of flat edges and/or flat triangles are found in a preprocessing stage, and they are tagged as maxima or non-maxima. The algorithm starts growing 2-cells from plateau maxima as well as for point maxima. A non-maximum plateau is assigned to the same 2-cell as the triangle t, which is edge-adjacent to the plateau, and has the highest third vertex. In addition, special cases of the inclusion rules are defined for the inclusion of triangles having one flat edge (see [15] for details).

An extension of this algorithm to three dimensions would need more vertex labels, and more cases to be considered for the inclusion of a new tetrahedron into the current 3-cell. An extension to higher dimensions would cause an explosion of the number of cases, and so seems not viable.

3.2.3 The Algorithm by Gyulassy et al.

The algorithm proposed by Gyulassy et al. [14] computes the Morse-Smale complex on a tetrahedral mesh Σ. The output of this algorithm represents the ascending and descending cells of all dimensions in the Morse complexes, where each cells is encoded as a collection of vertices of Σ.

The ascending cells are computed iteratively in order of decreasing dimension through a region-growing process. Descending cells are computed inside the ascending 3-cells using the same region-growing approach.

The computation of the ascending 3-cells consists of two steps:

1. The set of minima of f are identified; each minimum is used as the origin to build a set of vertices representing its ascending 3-cell, and they are processed in order of increasing field value.
2. Each vertex p of Σ is classified as an internal vertex of an ascending cell, or as a boundary vertex. This depends on the number of connected components of the set of internal vertices in the lower link of p which are already classified as interior to some ascending 3-cell. The classification is performed by sweeping Σ in order of ascending function values.

Vertices classified as boundary by the above algorithm are the input for the algorithm which builds the ascending 2-cells. An ascending 2-cell is created for each pair of adjacent 3-cells. The vertices of the 2-cells are classified as interior or boundary based on local neighborhood information, similarly to the classification done with respect to the 3-cells.

Then, the algorithm computes the ascending 1-cells. A 1-cell is created at every crossing of ascending 2-cells. Each 1-cell is composed of vertices classified as boundary in the previous step. Finally, each vertex p of an ascending 1-cell is classified as interior or boundary. Maxima are created at the boundaries between ascending 1-cells.

For each ascending 3-cell, the descending cells are computed in their interior. The region-growing steps are symmetric. Again here, the iteration is performed in the order of decreasing dimension.

Each vertex is examined by the algorithm at most three times. Under the assumption that the size of the link of each vertex is bounded by a constant, the overall time complexity is dominated by the cost of sorting, i.e., $O(m \log m)$, where m is the number of vertices in the tetrahedral mesh.

3.2.4   Analysis and Comparisons

The algorithms reviewed in this section are summarized in Table 3.2. The two methods in [5, 6] (see Sect. 3.2.1) use the same basic approach, but the former considers simply the difference of function values between two triangle vertices, while the latter considers the slope of triangles. This leads to a difference in their results, as illustrated in Sect. 3.1.2 for the case of boundary-based algorithms.

The criteria for including triangles used in [15] (see Sect. 3.2.2) are based just on vertex elevation, as in [6]. There is only one difference between the criteria for including a triangle used in [6, 15]. In [6], a candidate triangle t is never included if the third vertex of t is labeled T (i.e., in the situation of Fig. 3.10c and d). Instead, as long as unclassified triangles exist, [6] restarts from one vertex (not necessarily a maximum) that is labeled S in all its incident triangles that have not yet been classified. This is equivalent to examining fans in [15]. The real difference is that [6] assigns a whole fan to one maximum (the one with the highest elevation), while [15] divides the fan in correspondence of the lowest valley edge, which is a more intuitive behavior (see Fig. 3.11).

Unlike boundary-based methods, region-growing algorithms do not guarantee that saddle points lie on the 1-cells of the computed Morse complex, but they guarantee that the resulting 2-cells are uniformly two-dimensional, without dangling one-dimensional parts. For example, in Fig. 3.4b, the boundary edge chain of the

Table 3.2 Summary of reviewed region-growing algorithms

Algorithm	Input	Output	Region-growing process
Danovaro et al. [6]	2D/3D	Morse	Based on elevation difference, partial regions with final merging
Danovaro et al. [5]	2D/3D	Morse	Based on slope, partial regions with final merging
Magillo et al. [15]	2D	Morse	Based on elevation difference, entire regions
Gyulassy et al. [14]	3D	Morse-Smale	Iteratively compute ascending cells of dimension 3, 2, 1, and then descending cells inside ascending cells

For all algorithms, the input is a simplicial complex

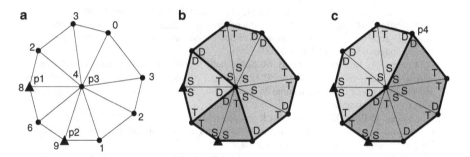

Fig. 3.11 (a) A triangle mesh with two maxima p_1 and p_2. (b) Danovaro et al. [6] build the 2-cells of vertices p_1, p_2, and p_3; later, the 2-cell of the last vertex (which is not a maximum) will be merged into the 2-cell of p_2. (c) Magillo et al. [15] further enlarge the 2-cell of each maximum; edge $p_3 p_4$ (lowest valley line) marks the boundary between the two 2-cells of p_1 and p_2

2-cell of maximum at elevation 9, found by boundary-based algorithms, connects vertices at elevations 2 (minimum), 5 (saddle), 3, 0 (minimum), 3, 6 (saddle), 1 (minimum), and the 2-cell includes the dangling edge (3,0). The 2-cell constructed by region-based algorithms is bounded by 2 (minimum), 5 (saddle), 3, 6 (saddle), 1 (minimum): one minimum is missing and there are two consecutive saddles, but there are no dangling edges.

In [16], it has been experimentally found that the boundary-based approach using elevation difference [21] and the region-growing approach in [15] give the same classification of triangles on a triangle mesh which does not contain flat edges. This result refers just to triangle classification, and does not consider the fact that boundary-based 2-cells may have dangling edges.

Region-growing methods either are designed with, or they can be easily equipped with ad-hoc solutions for dealing with flat edges. It is sufficient to identify plateaus in a preprocessing stage, and then to treat them as special cases both in the initialization and in the growing process of 2-cells. An analysis and experimental results about the problems caused by the presence of flat edges are presented in Sect. 7.4.

References

1. C. L. Bajaj, V. Pascucci, and D. R. Shikore. Visualization of scalar topology for structural enhancement. In *Proc. IEEE Visualization'98*, pages 51–58. IEEE Computer Society, 1998.
2. C. L. Bajaj and D. R. Shikore. Topology preserving data simplification with error bounds. *Computers and Graphics*, 22(1):3–12, 1998.
3. P.-T. Bremer, H. Edelsbrunner, B. Hamann, and V. Pascucci. A multi-resolution data structure for two-dimensional Morse functions. In *Proc. IEEE Visualization'03*, pages 139–146. IEEE Computer Society, October 2003.
4. P.-T. Bremer, H. Edelsbrunner, B. Hamann, and V. Pascucci. A topological hierarchy for functions on triangulated surfaces. *Transactions on Visualization and Computer Graphics*, 10(4):385–396, July/August 2004.

5. E. Danovaro, L. De Floriani, P. Magillo, M. M. Mesmoudi, and E. Puppo. Morphology-driven simplification and multiresolution modeling of terrains. In E.Hoel and P.Rigaux, editors, *Proc. ACM GIS 2003 - The 11th International Symposium on Advances in Geographic Information Systems*, pages 63–70. ACM Press, 2003.
6. E. Danovaro, L. De Floriani, and M. M. Mesmoudi. Topological analysis and characterization of discrete scalar fields. In T.Asano, R.Klette, and C.Ronse, editors, *Geometry, Morphology, and Computational Imaging*, volume 2616 of *Lecture Notes in Computer Science*, pages 386–402. Springer Verlag, 2003.
7. M. de Berg, P. Bose, K. Dobrindt, M.J. van Kreveld, M.H. Overmars, M. de Groot, T. Roos, J. Snoeyink, and S. Yu. The complexity of rivers in triangulated terrains. In *Canadian Conference on Computational Geometry*, pages 325–330, 1996.
8. M. de Berg, O. Cheong, H. Haverkort, J.G. Lim, and L. Toma. The complexity of flow on fat terrains and its i/o-efficient computation. *Computational Geometry: Theory and Applications*, 43:331–356, 2010. Special issue on the 10th Workshop on Algorithms and Data Structures (WADS).
9. M. de Berg and C. Tsirogiannis. Exact and approximate computations of watersheds on triangulated terrains. In *Proc. 19th ACM SIGSPATIAL International Conference on Advances in Geographic Information Systems*, pages 74–83, 2011.
10. H. Edelsbrunner and J. Harer. The persistent Morse complex segmentation of a 3-manifold. In Nadia Magnenat-Thalmann, editor, *3DPH*, volume 5903 of *Lecture Notes in Computer Science*, pages 36–50. Springer, 2009.
11. H. Edelsbrunner, J. Harer, V. Natarajan, and V. Pascucci. Morse-Smale complexes for piecewise linear 3-manifolds. In *Proc. 19th ACM Symposium on Computational Geometry*, pages 361–370, 2003.
12. H. Edelsbrunner, J. Harer, and A. Zomorodian. Hierarchical Morse complexes for piecewise linear 2-manifolds. In *Proc. 17th ACM Symposium on Computational Geometry*, pages 70–79, 2001.
13. H. Edelsbrunner and E. P. Mücke. Simulation of simplicity: a technique to cope with degenerate cases in geometric algorithms. *ACM Transactions on Graphics*, 9(1):66–104, 1990.
14. A. Gyulassy, V. Natarajan, V. Pascucci, and B. Hamann. Efficient computation of Morse-Smale complexes for three-dimensional scalar functions. *IEEE Transactions on Visualization and Computer Graphics*, 13(6):1440–1447, Nov-Dec 2007.
15. P. Magillo, E. Danovaro, L. De Floriani, L. Papaleo, and M. Vitali. A discrete approach to compute terrain morphology. *Computer Vision and Computer Graphics Theory and Applications*, 21:13–26, 2009.
16. P. Magillo, L. De Floriani, and F. Iuricich. Morphologically-aware elimination of flat edges from a tin. In *Proc. 21st ACM SIGSPATIAL International Conference on Advances in Geographic Information Systems (ACM SIGSPATIAL GIS 2013)*, November 5-8 2013.
17. V. Pascucci. Topology diagrams of scalar fields in scientific visualization. In S. Rana, editor, *Topological Data Structures for Surfaces*, pages 121–129. John Wiley & Sons Ltd, 2004.
18. B. Schneider. Extraction of hierarchical surface networks from bilinear surface patches. *Geographical Analysis*, 37(2):244–263, 2005.
19. B. Schneider and J. Wood. Construction of metric surface networks from raster-based DEMs. In S. Rana, editor, *Topological Data Structures for Surfaces*, pages 53–70. John Wiley & Sons Ltd, 2004.
20. S. Takahashi. Algorithms for extracting surface topology from digital elevation models. In S. Rana, editor, *Topological Data Structures for Surfaces*, pages 31–51. John Wiley & Sons Ltd, 2004.
21. S. Takahashi, T. Ikeda, T. L. Kunii, and M. Ueda. Algorithms for extracting correct critical points and constructing topological graphs from discrete geographic elevation data. In *Computer Graphics Forum*, volume 14, pages 181–192, 1995.

Chapter 4
Watershed Algorithms

The watershed approach has been developed in image processing for segmentation of grey-level images, which can be viewed as 2D scalar fields modeled as regular grids. It extends directly to higher dimensions, for instance to 3D scalar fields. The theoretical counterpart for smooth functions has been presented in Sect. 1.4.

The intuitive idea is computing the ascending Morse complex of the scalar field by simulating the diffusion of water, where the scalar field is considered as a height field (e.g., a terrain). A first approach is based on the idea of flooding the terrain. This is somehow a region-based approach, in the sense that it takes the minima as seeds, and progressively grows the 2-cells of the ascending Morse complex (called catchment basins) from them. The approach based on topographic distance does a similar thing by computing shortest paths from minima. Another approach is based on the opposite idea of simulating the descent of water from any point of the terrain down to a minimum.

Most watershed algorithms work on d-dimensional regular grids, where the field values are associated with d-cells. The adjacency relations between d-cells can be modeled in two ways: each d-cell γ is considered as adjacent either to the d-cells which share a $(d-1)$-face with it, or to all the d-cells sharing a face of any dimension with γ (that is, 4- and 8-connectivity for square grids, see Sect. 1.2).

The output is a classification of the d-cells of the grid as belonging to the catchment basin (ascending d-manifold) of a certain minimum, or as belonging to the boundary between catchment basins.

Few watershed algorithms are defined for simplicial models. In that case, they work on the primal graph of the model and their output is a classification of the vertices as belonging to the catchment basin (ascending d-manifold) of a certain minimum, or as belonging to the boundary between catchment basins.

In general, we can say that watershed algorithms operate on a labeled graph $G = (V, E, f)$, where V is the set of reference cells (d-cells for a grid, vertices for a simplicial model), E describes the adjacency relation, and function $f : V \longrightarrow \mathbb{R}$ is the field value. Similarly, the output of such algorithms can be considered as a

© The Author(s) 2014
L. Čomić et al., *Morphological Modeling of Terrains and Volume Data*,
SpringerBriefs in Computer Science, DOI 10.1007/978-1-4939-2149-2_4

labeling of the nodes of the graph, i.e., a function $lab : V \longrightarrow V \cup \{\text{watershed}\}$ that associates each node u with the minimum node of the catchment basin containing u, or with the special value denoting that u is a watershed node. Not all watershed approaches, however, produce watershed labels.

In Sects. 4.1–4.3, respectively, we present algorithms belonging to the three approaches to watershed computation, i.e., by simulating water raising from minima, by topographic distance, and by simulating water descent to minima. In Sect. 4.4 we compare the various approaches.

4.1   Watershed by Simulated Immersion

The intuition behind the simulated immersion approach is the following. On a terrain, we drill holes in place of local minima, immerse this terrain in a pool, let water raise from drilled holes, and build dams to prevent water coming from different minima to merge. Then, the watershed of the terrain is described by these dams (labeled as watershed nodes), and the catchment basins are delineated by the dams.

More formally, simulated immersion relies on a definition of the watershed transform in the discrete domain, where the catchment basins are defined recursively, using the concept of skeleton by influence zones [1]. The following definitions refer to the input scalar field as a labeled graph $G = (V, E, f)$.

Definition 17. Given a graph $G = (V, E, f)$ and two nodes $a, b \in V$, the *geodesic distance* $d(a, b)$ between a and b is the length of the shortest path connecting a and b in G.

In the above definition, the length of a path is computed in co-domain space (i.e., in 3D for a terrain model).

The geodesic distance of a node a from a set of nodes $B \subseteq V$ is the minimum geodesic distance between a and nodes $b \in B$.

Given k disjoint subsets of nodes $B_i \subseteq V$ ($i = 1, \ldots k$), where each B_i forms a connected subgraph of G, the *geodesic influence zone* of B_i is defined as the set of all nodes $a \in V$ such that a is closer to B_i than to any other B_j with $j \neq i$:

Definition 18. The *geodesic influence zone* of B_i is:

$$IZ(B_i) = \{a \in V \mid d(a, B_i) < d(a, B_j) \text{ for all } j \neq i\}$$

The *simulated immersion* algorithm presented by Vincent and Soille [12, 14] expands catchment basins by processing the nodes of G by increasing elevation. In the stage in which a certain elevation h is processed, all catchment basins of minima at elevations $h' < h$ have been started and, up to now, contain just nodes at elevations lower than h. Processing elevation h will add new nodes, lying at elevation h, to existing basins, and will start new basins from minima having an elevation equal to h.

A recursive definition of catchment basins is used. Let $h_1, \ldots, h_k$ be the ordered set of values assumed by function f. The union X_i of all catchment basins up to elevation h_i (and its complement $W_i = V \setminus X_i$ of watershed nodes up to elevation h_i) is incrementally constructed, where X_k is the final watershed of graph G.

Definition 19. Set $X_h \subseteq V$ is recursively defined as follows:

- For the lowest function value $h = h_1$, X_1 contains the minima at elevation h_1 (absolute minima of the function).
- For any subsequent function value $h = h_i$, X_i contains the minima at elevation h_i, plus all the nodes v which have $f(v) \leq h_i$ and belong to the geodesic influence zone $IZ(X_{i-1})$ of X_{i-1}.

The two terms in the construction of X_i correspond to initialization of new basins from seeds at elevation h_i, and to the expansion of existing basins by including adjacent nodes at elevation h_i. Minima can be isolated nodes of G, or plateaus.

The algorithm performs the following steps:

- Sort nodes by increasing elevation.
- Process elevations, starting from the lowest one and ending at the highest one. For the current elevation h:

  1. Initialize the label $lab(v)$ of each node v at elevation h with a special mask value (meaning "to be set").
  2. Initialize a FIFO (First In-First Out) queue Q containing all nodes v at elevation h, which have some neighbor already labeled with a basin.
  3. While queue Q is not empty, extract a node v and do the following: if all neighbors of v, which are labeled with a basin, have the same basin label b, then set $lab(v) = b$; otherwise (i.e., there are more basins among the neighbors of v), then set $lab(v) =$ watershed. Then, insert into the queue all neighbors of v that are labeled with the mask.

     When the queue is empty, all nodes at elevation h that could be reached from an existing basin (with seed at a lower elevation than h) have been labeled. Now, basins whose seed is at elevation h have to be initialized.
  4. Start another loop over nodes v at elevation h. If v is still marked with the mask, then set $lab(v)$ to a new basin label b (v is a seed of a new basin), and assign label b to all nodes of the same connected component of nodes at elevation h, which are labeled with the mask (found through a breadth first search from v).

The FIFO management of queue Q guarantees that nodes belonging to the same (non-minimum) plateau at elevation h are processed in order of increasing distance from the boundary of the plateau. This permits to place watershed pixels near the middle of the plateau, in case the plateau must be split among different catchment basins.

This algorithm has been originally defined on grey-level images (2D scalar fields), and it extends to 3D scalar fields in a straightforward way. It has also been extended to triangle meshes in [14]. A version for simplicial models (using the primal graph and generating a vertex classification), both in 2D and in 3D, has been developed in [6, 7].

The computational complexity includes the cost for the preliminary sorting of elevations, which is in $O(m \log m)$ as the m nodes of the graph may have m distinct elevations. The main loop of the algorithm has a cost in $O(m + e)$ where m and e denote the number of nodes and arcs of graph G, respectively. If G corresponds to a 2D scalar field (either modeled as a square grid, or as a triangle mesh), then $e = O(m)$ and the total cost is thus in $O(m \log m)$. If G corresponds to a 3D scalar field, then $e = O(m)$ for a grid (each cubic cell is connected to a fixed number of adjacent cells). In practice, it is linear in m for a tetrahedral mesh as well, although it can be $e = O(m^2)$ in the worst case.

4.2 Watershed by Topographic Distance

This approach, due to Meyer [10], tries to mimic the definition of catchment basins given in the smooth case, by referring to a discrete version of the topographic distance.

First, we define the lower slope at node p, which is the maximal slope linking p to any of its neighbors of lower elevation.

Definition 20. Given a graph $G = (V, E, f)$ and a node $p \in V$, the *lower slope* $LS(p)$ is defined as

$$
LS(p) = \begin{cases} 0 & \text{if no neighbor } p' \text{ exists with } f(p') < f(p) \\ \max\{\frac{f(p)-f(p')}{dist(p,p')} \mid (p, p') \in E, f(p') < f(p)\} & \text{otherwise} \end{cases}
$$

In the above definition, distance $dist(p, p')$ is computed in domain space (i.e., on the 2D plane in case of a terrain).

The *cost* for walking from node p to an adjacent node p', i.e., the cost for traversing the directed arc (p, p'), is given by:

$$
cost(p, p') = \begin{cases} LS(p) \, dist(p, p') & \text{if } f(p) > f(p') \\ LS(p') \, dist(p, p') & \text{if } f(p) < f(p') \\ \frac{LS(p)+LS(p')}{2} \, dist(p, p') & \text{if } f(p) = f(p') \end{cases}
$$

In practice, the above cost function considers the actual elevation difference for the steepest edge, while it "stretches" other edges by a factor which gets larger if their slope is smaller.

Given a path $\pi = (p = p_0, \ldots, p_l = q)$ between two nodes p and q in the graph G, the π- *topographic distance* $TD^\pi(p, q)$ *of path* π is given by the sum of costs for traversing all directed arcs (p_i, p_{i+1}):

$$T^{\pi}(p,q) = \sum_{i=0}^{l-1} cost(p_i, p_{i+1}).$$

Definition 21. The *topographic distance* $T(p,q)$ between p and q is the minimum of the topographic distances along all paths between p and q:

$$T(p,q) = \min_{\pi=(p,\dots,q)} T^{\pi}(p,q).$$

For any path π from p to q, we have that $T^{\pi}(p,q) \geq |f(p) - f(q)|$, and the equality holds if and only if π is a path of steepest descent.

The topographic distance is not a true distance function because it is zero on distinct nodes, if they belong to the same plateau.

The definition of a catchment basin in a graph is similar to the definition given in the smooth case.

Definition 22. Let $\{m_i\}_{i \in I}$ be the set of minima of function f, where each m_i can be a single node or a plateau.
The *catchment basin* $CB(m_i)$ of a minimum m_i is defined by:

$$CB(m_i) = \{p \in D : f(m_i) + T(p,m_i) < f(m_j) + T(p,m_j), j \in I - \{i\}\}.$$

The *image integration* algorithm by Meyer [10, 11] is a variation of the Dijkstra-Moore algorithm [4] for computing the shortest path from a source node to every other node within a graph. In this case, topographic distance is used.

Dijkstra-Moore algorithm is incremental. For each node, there exists a current (tentative) shortest path, that is possibly finalized. The distance along the current (finalized or tentative) shortest path, and its previous node in such path are recorded. The algorithm uses a priority queue sorted by increasing distance.

The *image integration* algorithm [10] computes all shortest paths from each minimum simultaneously, and records, for every node, its closest minimum instead of the previous node along the path (since we are not interested in reconstructing the path). The idea is that of a moving wavefront. If two or more wavefronts reach the same node, we compare the two path lengths and assign the node to the closest minimum.

The algorithm stores two items associated with each node v: its current (tentative) topographic distance $td(v)$, and its source node $lab(v)$. It also uses a priority queue based on $td(v)$. The main steps of the algorithm are:

1. Compute the minima (including plateau minima).
2. For each minimum v, set $td(v) = f(v)$, and $lab(v) = v$ (if v belongs to a plateau, use a representative vertex of the plateau). If v has no adjacent vertices at a higher elevation (i.e., v lies in the interior of a plateau minimum), then finalize v; otherwise, insert v in a priority queue Q.
3. For each other node v, set $td(v)$ to infinity and $lab(v)$ as undefined.

4. While the priority queue Q is not empty, pop the first node v from Q (it is the one with minimum distance), and do the following:

   a. Finalize v.
   b. For each adjacent node u of v, such that u is not finalized, update its information as follows: let $d = td(v) + cost(u, v)$; if $d > td(u)$ then set $td(u) = d$ and $lab(u) = lab(v)$.
   c. If u is in the priority queue Q, update its position based on its new priority. Otherwise, insert u into Q.

The algorithm accepts plateaus. When the line separating two catchment basins should run through a (non-minimum) plateau, then the line drawn by the algorithm depends on the order in which nodes belonging to the plateau are picked from the queue, and, in general, is not the intuitively best one (i.e., equidistant from the boundaries of the plateau). In order to avoid this problem, nodes belonging to a plateau are further sorted, in a priority queue, based on their distance from the boundary of the plateau.

The *hill climbing* algorithm [10] is a simpler and faster version of the *image integration* algorithm, which applies to grids, since the distance $dist(p, p')$ between two adjacent pixels p and p' in domain space is constant. Hill climbing does not compute $td(v)$, but inserts a node v into the priority queue based on the value of $f(v)$. Step (b) in the previous algorithm is simply replaced with:

   b. For each adjacent node u of v, whose label is still undefined (this implies $f(v) < f(u)$), set $lab(u) = lab(v)$.

The hill climbing method replaces topographic distance with elevation difference between nodes. This gives the same result only on regular grids.

The worst-case computational complexity of these approaches is the same as that of simulated immersion.

The presented algorithms have been originally defined in 2D, and extend to higher dimensional scalar fields in a straightforward way.

4.3   Watershed by Rain Falling Simulation

All approaches presented so far have in common the idea of starting from the minima and letting the catchment basins grow until all nodes have been classified into catchment basins (or as watershed nodes). The approaches presented in [9, 13] use the opposite idea, called *rain falling paradigm*, where the steepest descending path is constructed from each node until it arrives either at a minimum, or at a node which has already been inserted in a catchment basin. The label of each minimum is propagated backwards along the steepest path. The result of this process is a segmentation of the nodes of graph G into catchment basins associated with the minima of f. No watershed nodes are generated.

The algorithm proposed by Mangan and Whitaker [9] has been designed for triangle meshes. The one by Stoev and Strasser [13] for regular grids.

The main steps of the two algorithms are:

1. Find the minima and label each minimum as belonging to the basin of itself.
2. For each node u, which is still unlabeled, perform a descent from u:

   - find the lowest neighbor v of u;
   - if v is labeled, give the label of v to u and, through backtracking, to all previously traversed nodes during the descent;
   - otherwise, recursively continue the descent from v (until it ends into a labeled node).

Possible issues with the rain falling paradigm are the non-uniqueness of the lowest neighbor q of a node p, and the occurrence of plateaus.

If a node p has more than one lowest neighbor, the algorithm in [9] does not explicitly show how to solve this ambiguity, while the one in [13] tries to continue the path ahead from all candidate lowest neighbors, and chooses the one that would go lowest.

Both methods have been designed for graphs which may contain plateaus, and handle them directly. In [9], all plateaus are found in a preliminary stage, and each of them is treated as a single node. Plateaus that are minima are processed in Step 1 by labeling all nodes of the plateau with the same basin label. In Step 2, plateaus which are not minima are processed before all other vertices. From each such plateau, a descent to its adjacent node of lowest elevation is performed.

In [13], plateaus are found during the descent. Step 1 considers just minima located at isolated nodes. In Step 2, when processing a node u having some neighbor with equal elevation, the plateau containing u is computed and its boundary is checked. If the plateau is a minimum, then a new basin label is given to that plateau and propagated backwards to the previous nodes of the path. If the plateau is not a minimum, then the descent continues to the lowest node adjacent to its boundary. If this node is not unique, the plateau is split into parts.

In [9], each non-minimum plateau is assigned entirely to a unique catchment basin, while in [13], a non-minimal plateau can be split among different catchment basins.

The advantage of the rain-falling approach, over the approaches based on flooding or topographic distance, is that it does not require a preliminary sorting of vertices, nor a priority queue. Therefore, these algorithms do not have an $O(m \log m)$ term in their computational complexity.

Also this method was first developed for grids, but it extends easily to simplicial models (by considering the primal graph and producing a vertex classification). An implementation for triangle meshes has been used in [8], where triangles are then classified based on the labels of their vertices. A similar approach can also be applied for tetrahedral meshes [7].

4.4 Summary and Comparisons

Table 4.1 presents a summary of reviewed algorithms. All watershed algorithms are dimension-independent.

Simulated immersion and rain falling simulation consider just the elevations and are independent on the shape of the underlying regular grid or simplicial model. The image integration method considers the length of edges as well, and thus it is somehow comparable to approaches for triangle meshes based on the discrete gradient, such as [2, 3, 5] (see Chap. 3). On regular grids with the 4-connectivity model, using the topographic distance (or gradient simulation) is equivalent to using the elevation difference.

Rain falling algorithms construct catchment basins by moving downwards, while the other two approaches do it by going upwards. We examine how such two choices lead to different results.

We have highlighted the problem on a small triangle mesh with 11 vertices and 11 triangles, having two minima in vertices v_0 and v_1 (see Fig. 4.1). Here, letter z denotes elevation. The rain falling and simulated immersion algorithms give two different vertex classifications.

Table 4.1 Summary of reviewed watershed algorithms

Algorithm	Input	Approach
Vincent and Soille [12, 14]	Grid (simplicial complex)	Simulated immersion
Meyer [10] (image integration)	Grid (simplicial complex)	Topographic distance
Meyer [10] (hill climbing)	Grid	Topographic distance
Mangan and Whitaker [9]	Triangle mesh (grid)	Rain falling
Stoev and Strasser [13]	Grid (simplicial complex)	Rain falling

We report the input format considered in the original definition of the algorithm, and (in brackets) the acceptable input

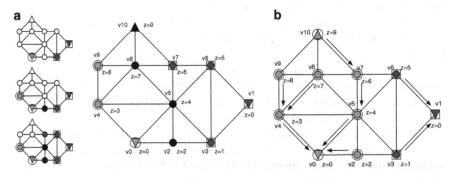

Fig. 4.1 Executing the two algorithms on a toy example: (**a**) simulated immersion, (**b**) rain falling. *Circled yellow* and *squared cyan* nodes correspond to the basins of minima v_0 and v_1, respectively. In (**a**), black nodes are watershed. (Color figure online)

Figure 4.1a shows three stages of the *simulated immersion* process, and the final vertex labeling. The three stages are: after processing $z = 1$ and just before labeling v_2 ($z = 2$) as watershed; after processing $z = 3$ and just before labeling v_5 ($z = 4$) as watershed; after processing $z = 6$. Vertices v_2, v_5, v_8, v_{10} are labeled as watershed vertices, and vertex v_7 ($z = 6$) is classified as belonging to the basin of v_1 because, when water reaches it, its labeled neighbor vertices are v_5 (watershed) and v_6 (whose basin is v_1).

Figure 4.1b shows the steepest descent from each vertex, and the final vertex labeling produced by the *rain falling* algorithm. The steepest descent from v_7 ($z = 6$) is towards v_5 ($z = 4$), and from v_5 to minimum v_0. Therefore, vertex v_7 is labeled with the basin of v_0 (whereas it is labeled v_1 by the simulated immersion algorithm).

Fig. 4.2 Lines of steepest ascent (*dashed red*) and descent (*plain blue*) through vertex v_7 on the toy example. (Color figure online)

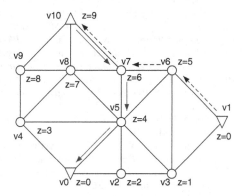

The key reason why we have two different vertex classifications is that the simulated immersion algorithm classifies a vertex v by taking into account which minimum m first reaches v when water raises from m, while the rain falling algorithm classifies v based on which minimum m' is first reached when water falls from v. The two perspectives are not equivalent on a discretized terrain, because the line of steepest ascent through a vertex v may not be the same as the line of steepest descent through v. In the continuous case, the two lines are the same. On our toy example, the line of steepest ascent through vertex v_7 is $v_1 \rightarrow v_6 \rightarrow v_7 \rightarrow v_{10}$ and the line of steepest descent through v_7 is $v_{10} \rightarrow v_7 \rightarrow v_5 \rightarrow v_0$ (see Fig. 4.2). Therefore, depending on the approach used, vertex v_7 is assigned to minimum v_1 or v_0.

References

1. S. Beucher and C. Lantuejoul. Use of watersheds in contour detection. In *International Workshop on Image Processing: Real-Time Edge and Motion Detection/Estimation, Rennes, France*, September 17-21, 1979.

2. P.-T. Bremer, H. Edelsbrunner, B. Hamann, and V. Pascucci. A multi-resolution data structure for two-dimensional Morse functions. In *Proc. IEEE Visualization'03*, pages 139–146. IEEE Computer Society, October 2003.
3. E. Danovaro, L. De Floriani, P. Magillo, M. M. Mesmoudi, and E. Puppo. Morphology-driven simplification and multiresolution modeling of terrains. In E.Hoel and P.Rigaux, editors, *Proc. ACM GIS 2003 - The 11th International Symposium on Advances in Geographic Information Systems*, pages 63–70. ACM Press, 2003.
4. E. W. Dijkstra. A note on two problems in connexion with graphs. *Numerische Mathematik*, 1:269–271, 1959.
5. H. Edelsbrunner, J. Harer, and A. Zomorodian. Hierarchical Morse complexes for piecewise linear 2-manifolds. In *Proc. 17th ACM Symposium on Computational Geometry*, pages 70–79, 2001.
6. L. De Floriani, F. Iuricich, P. Magillo, and P. D. Simari. Discrete Morse versus watershed decompositions of tessellated manifolds. In *ICIAP (2)*, pages 339–348, 2013.
7. F. Iuricich. *Multi-resolution shape analysis based on discrete Morse decompositions*. PhD thesis, University of Genova – DIBRIS, Italy, 2014.
8. P. Magillo, L. De Floriani, and F. Iuricich. Morphologically-aware elimination of flat edges from a tin. In *Proc. 21st ACM SIGSPATIAL International Conference on Advances in Geographic Information Systems (ACM SIGSPATIAL GIS 2013)*, November 5-8 2013.
9. A. Mangan and R. Whitaker. Partitioning 3D surface meshes using watershed segmentation. *Transactions on Visualization and Computer Graphics*, 5(4):308–321, 1999.
10. F. Meyer. Topographic distance and watershed lines. *Signal Processing*, 38:113–125, 1994.
11. F. Meyer and S. Beucher. Morphological segmentation. *J. of Visual Communication and Image Representation*, 1:21–45, 1990.
12. P. Soille. *Morphological Image Analysis: Principles and Applications*. Springer-Verlag, Berlin and New York, 2004.
13. S. L. Stoev and W. Strasser. Extracting regions of interest applying a local watershed transformation. In *Proc. IEEE Visualization'00*, pages 21–28. ACM Press, 2000.
14. L. Vincent and P. Soille. Watershed in digital spaces: An efficient algorithm based on immersion simulation. *IEEE Transactions on Pattern Analysis and Machine Intelligence*, 13(6):583–598, 1991.

Chapter 5
A Combinatorial Approach Based on Forman Theory

In this chapter, we review algorithms based on Forman theory. These algorithms are combinatorial in nature and dimension-independent. They can be applied to both 2D and 3D scalar fields. We refer to Sect. 1.6 for the background notions on Forman theory.

The input scalar field f is defined at the vertices of a cell complex. All algorithms based on Forman theory include a preliminary step which, starting from f, builds a *Forman function* F defined on all the cells of the complex, or directly a Forman gradient vector field V (see Sect. 1.6). Given a field V, they then produce the descending and ascending Morse complexes, and the Morse-Smale complex. Such complexes are called *discrete Morse* and *discrete Morse-Smale complexes*.

In Sect. 5.1 we describe an encoding for Morse and Morse-Smale complexes on a cell complex, and, in more detail, for triangle and tetrahedral meshes. The primal/dual relationship between the descending and ascending Morse complexes has been expressed in terms of the *primal* simplicial complex Σ, and its *dual* complex Σ^* (see Sect. 1.1). The combinatorial structure of the Morse-Smale complex is expressed as a collection of cells in the complex obtained by the intersection of Σ and Σ^*, which we refer to as the *dually subdivided* mesh Σ_S. According to the primal/dual representation, these complexes can be fully described in terms of vertices and d-simplices of the complex. In Sect. 5.2 we describe the encoding for a Forman gradient defined over simplicial complexes proposed in [17]. Such representation stores information only at maximal simplices, and has been used as a compact representation for the Forman gradient of a Forman function defined over a simplicial complex.

In Sect. 5.3, we describe the most important algorithms in the literature for computing a Forman gradient on 2D and 3D scalar fields. In Sect. 5.4, we present existing approaches which, given a Forman gradient F, extract the collection of cells of the ascending and descending Morse complex, the Morse-Smale cells, or the 1-skeleton of the Morse-Smale complex.

© The Author(s) 2014
L. Čomić et al., *Morphological Modeling of Terrains and Volume Data*,
SpringerBriefs in Computer Science, DOI 10.1007/978-1-4939-2149-2_5

5.1   Representing Morse Complexes in the Discrete Case

Representations for descending and ascending Morse complexes can be defined independently of the representation used to discretize the domain of the Morse function. However in application domains simplicial complexes and regular grids are used in order to optimize both the computation and the analysis of such complexes. As described in [10], computing the Morse and Morse-Smale complexes on a cell complex Γ means to group the i-cells, with $0 \leq i \leq d$, forming all the ascending and descending Morse cells or the cells of the Morse-Smale complex. In [17] a new interpretation of the Morse and Morse-Smale complexes in terms of the input cell complex and its dual complex has been presented.

5.1.1   Representing Discrete Ascending and Descending Morse Complexes

Let us consider a simplicial complex or a regular grid as underlying decompositions of a scalar field model. We first observe some properties of the ascending and descending Morse complexes of a Forman function F (and the corresponding Forman gradient field V). As described in Sect. 1.6, in the 2D case, a *descending 2-cell* corresponds to a maximum, and thus to a collection of primal 2-cells (triangles or squares) in the underlying simplicial complex or regular grid. An *ascending 2-cell* corresponds to a minimum, and thus to a collection of dual 2-cells, each of which corresponds to a primal vertex. A *descending 1-cell* corresponds to a saddle, and thus to a sequence of primal edges. An *ascending 1-cell* corresponds to a saddle as well, and thus to a sequence of dual edges, each of which corresponds to a primal edge. Therefore, the descending Morse complex consists of elements from the primal complex, while the ascending Morse complex consists of elements from the dual complex.

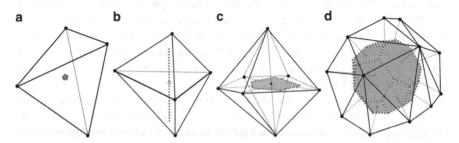

Fig. 5.1 The primal/dual relationships in a tetrahedral mesh. (**a**) The dual of a tetrahedron is a vertex, (**b**) the dual of a triangle shared by two tetrahedra is an edge (*dotted line*), (**c**) the dual of an edge is a polygon and (**d**) the dual of a vertex is a 3-cell

In the 3D case:

- A *descending 3-cell* corresponds to a maximum, and, thus, to a collection of (primal) 3-cells (see Fig. 5.1a). Dually, an *ascending 3-cell* corresponds to a minimum, and thus to a collection of dual 3-cells (i.e., primal vertices) (see Fig. 5.1d).
- A *descending 2-cell* corresponds to a 2-saddle, and, thus, to a collection of primal 2-cells (see Fig. 5.1b), each of which can be expressed as a pair of primal 3-cells. An *ascending 2-cell* corresponds to a 1-saddle, and thus to a collection of dual 2-cells (see Fig. 5.1c), each of which can be expressed as a pair of dual 3-cells, corresponding to a primal edge.
- A *descending 1-cell* corresponds to a 1-saddle and, thus, to a sequence of primal edges (see Fig. 5.1c), or, equivalently, to a sequence of primal vertices. An *ascending 1-cell* corresponds to a 2-saddle and thus to a sequence of dual edges (see Fig. 5.1b), which can be seen as a pair of dual vertices (i.e., as a sequence of primal 3-cells).

Note that, since the primal edges can be expressed as pairs of primal vertices and the primal faces can be expressed as pairs of primal maximal cells, all the Morse cells can be expressed in terms of primal vertices and primal maximal cells of the underlying complex.

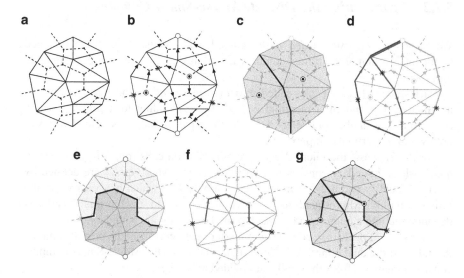

Fig. 5.2 Two-dimensional example of the scheme for simplicial complexes. (**a**) The triangle mesh Σ (*solid lines*) is overlapped with the dual complex Σ^* (*dashed lines*). (**b**) Encoding the Forman gradient field entirely with the triangles enables the use of compact topological data structures for morphological extraction. The cells of the *descending Morse complex* are associated with the cells of Σ ((**c**)–(**d**)) and those of the *ascending Morse complex* with the cells of Σ^* ((**e**)–(**f**)). Finally, the *Morse-Smale complex* is associated with the *dually subdivided* tetrahedral mesh Σ_S (**g**) whose quad-cells are defined by a triangle and one of its vertices. All relations are encoded strictly in terms of the vertices and triangles of Σ

Figure 5.2 illustrates the above observations for a Forman gradient field defined on a triangle mesh Σ, where the descending 2-cells (see Fig. 5.2c) are collections of triangles from Σ associated with the maxima (red critical points), while the ascending 2-cells (see Fig. 5.2e) are collections of dual 2-cells (corresponding to vertices from Σ) associated with the minima (blue critical points). Similarly, the descending (see Fig. 5.2d) and ascending 1-manifolds (see Fig. 5.2e) correspond to collections of primal and dual edges, respectively, associated with the saddles (green critical points).

Using the above correspondence, it can be noted that the i-saddles of the Forman gradient correspond to i-cells in the primal complex Σ and to $(d - i)$-cells in the dual complex Σ^*. Equivalently a *descending i-cell* corresponds to an i-saddle, thus it corresponds to a collection of primal i-cells or, equivalently, to a collection of dual $(d - i)$-cells. Dually, an *ascending i-cell* corresponds to a $(d - i)$-saddle, and thus to a collection of dual i-cells or, equivalently, to a collection of primal $(d - i)$-cells.

Note that all the descending and ascending manifolds are expressed entirely in terms of d-cells and vertices. This is a relevant issue from an implementation point of view, since there is no need to encode the primal i-cells (where $0 < i < d$), and, of course, no need for encoding the dual complex.

5.1.2   Representing the Discrete Morse-Smale Complex

We define the *dually subdivided mesh*, denoted as Σ_S, as the cell complex obtained by the intersection of the primal complex Σ and its dual Σ^* (see Fig. 5.2a for a 2D example).

Each maximal cell of Σ_S, called maximal *micro-cell*, is the intersection of a maximal cell γ and of a maximal cell γ^* (corresponding to a primal vertex). A micro-d-cell can be completely described by a pair of indices $<d$-cell,vertex$>$, both within the primal complex.

For the 2D case, the micro-2-cells are quadrilateral cells (quads) called *micro-quads* whose boundary consists of four *micro edges*. Micro-quads are defined by a primal 2-cell γ and a 2-cell in the dual complex Σ^* (which is a vertex v in the primal complex). Micro-quads are encoded as a pair (γ, v). Each primal 2-cell γ is decomposed into a micro-quad for each primal vertex on the boundary of γ.

If the cells of the Morse complexes consist of elements from the primal complex Σ and from the dual complex Σ^*, the (macro) cells of the Morse-Smale complex consist of elements from the dually subdivided mesh Σ_S.

In the 2D simplicial case: a Morse-Smale 2-cell is a collection of micro-quads; a Morse-Smale 1-cell is a collection of micro-edges on the boundary of two micro-quads. Figure 5.2g illustrates the Morse-Smale complex associated with the Forman gradient field of Fig. 5.2b. Note that each micro-quad is defined by the intersection of a triangle (a primal 2-cell) and a dual 2-cell associated with one of its boundary vertices, and that each (macro) 2-cell of the Morse-Smale complex is defined by a maximum (red critical point), a minimum (blue critical point) and two saddles (green critical points).

5.2 Encoding the Forman Gradient

According to the primal/dual representation for discrete Morse and Morse-Smale complexes described in Sect. 5.1 on a d-dimensional simplicial complex or regular grid Γ, these complexes can be fully described in terms of vertices and d-cells of Γ. On the contrary, for encoding a Forman gradient which is defined on all the cells of Γ, a natural representation for Γ would be the *incidence graph*.

An *Incidence Graph (IG)* [4] is a topological data structure encoding explicitly all the cells of a cell complex Γ and all the incidence relations among such cells. For each i-cell γ, it stores:

– the relation with the $(i - 1)$-cells in the combinatorial boundary of γ, and
– the relation with the $(i + 1)$-cells in the star of γ.

If we consider as discrete model a cubic grid, we can implicitly represent all such boundary/co-boundary relations among its cells. Due to its regularity, all relations can be represented by indexing the 3D cells of the grid. Moreover, since a Forman gradient V defines a pairing between incident cells, V can be defined on such representation as a bit vector based on the same indexing [5].

When encoding simplicial complexes, the incidence graph can be verbose, since we would need to explicitly encode all simplices in the complex plus the incidence relations above. Data structures, which encode only vertices and d-simplices [6, 13], have been shown to be much more compact [1]. Thus, a new encoding for a Forman gradient defined over simplicial complexes has been defined in [17], called a *compact gradient*.

5.2.1 Encoding Triangle and Tetrahedral Meshes

The underlying simplicial complex Σ is represented through an *Indexed data structure with Adjacencies (IA)* introduced in [12]. The IA data structure is a dimension-independent data structure encoding the 0- and d-simplices of Σ explicitly, plus the following relations:

- for each d-simplex σ:

  – the $d + 1$ vertices of σ;
  – the $d + 1$ d-simplices which share a $(d - 1)$-simplex with σ;

- for each 0-simplex v:

  – the n coordinates of v plus its fields value, where $n \geq d$ is the dimension of the embedding space.
  – one d-simplex incident in v;

All the 0-simplices (vertices) of Σ are stored in an array Σ_0. Similarly, all the d-simplices of Σ are stored in an array Σ_d. Note that each vertex v of a d-simplex σ defines a unique $(d-1)$-face of σ, which is the one that does not contain v.

5.2.2   Compact Gradient Encoding

When using the IA data structure for a triangle or a tetrahedral mesh, information regarding the Forman gradient V are attached only to the d-simplices [10, 17]. Let us consider a d-dimensional simplicial complex Σ and a d-simplex σ in Σ. The encoding associates with σ a subset of the pairs involving its faces. Specifically, it encodes all pairs of the discrete vector field of the form:

- (τ_i, τ_j), with τ_i, τ_j belonging the star $*\sigma$ of simplex σ;
- (τ_i, σ'), where σ' is one of the d-simplices adjacent to σ and τ_i is the common $(d-1)$-face between σ and σ'.

In a d-dimensional simplicial complex Σ, a d-simplex σ has $\binom{d+1}{i+1}$ faces of dimension i, and each face has in turn $(i+1)$ faces of dimension $(i-1)$. Since each i-simplex can be paired with any of the simplices on its boundary or co-boundary, there are $\sum_{i=1}^{d-1} \binom{d+1}{i+1} \cdot (i+1)$ possible pairs in the restriction of the Forman gradient V to σ. Adding the $d+1$ additional pairs from a $(d-1)$-simplex on the boundary of σ to an adjacent d-simplex provides a total of $\sum_{i=1}^{d-1} \binom{d+1}{i+1} \cdot (i+1) + d + 1$ possible pairs.

Let us consider a triangle mesh Σ as an example. The encoding associates with a triangle (2-simplex) σ in Σ a subset of the pairs involving its faces. Specifically, σ encodes all the pairs:

- (τ_1, σ'), corresponding to an arrow from an edge τ_1 to a triangle σ' (dotted arrows in Fig. 5.3);
- (τ_0, τ_1), corresponding to an arrow from a vertex τ_0 to an edge τ_1 (bold arrows in Fig. 5.3).

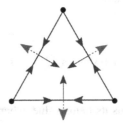

Fig. 5.3 Set of arrows per triangle σ. *Bold blue arrows* indicate pairs involving simplices belonging to the boundary of σ (and possibly σ itself). *Dotted red arrows* indicate pairs involving the edges of σ and the adjacent triangles of σ. (Color figure online)

Then, in a triangle mesh, a triangle has $\binom{3}{i+1}$ faces of dimension i, and each face has $(i + 1)$ simplices of dimension $(i - 1)$ on its boundary. Thus, there are

$$\sum_{i=1}^{2} \binom{3}{i+1} \cdot (i + 1) = 3 \cdot 2 + 1 \cdot 3 = 9$$

possible pairs in the restriction of vector field V to σ. Adding the three additional pairs from an edge of σ to an adjacent triangle gives a total of 12 possible pairs.

Similarly, a tetrahedron σ in a tetrahedral mesh has $\binom{4}{i+1}$ faces of dimension i, and

$$\sum_{i=1}^{3} \binom{4}{i+1} \cdot (i + 1) + 4 = 6 \cdot 2 + 4 \cdot 3 + 1 \cdot 4 + 4 = 32$$

possible pairs.

Such collection of pairs from the Forman gradient V in the vicinity of a maximal simplex σ is referred to as a *local frame* of the Forman gradient. Since each such pair within a local frame encodes a single bit of information (i.e. the presence or absence of that pair), each local frame can be encoded using $\sum_{i=1}^{d-1} \binom{d+1}{i+1} \cdot (i + 1) + d + 1$ bit flags per d-simplex. This bit flag representation simplifies testing for the presence of pairs as well as updates to the discrete vector field.

The restrictions imposed by discrete vector fields (i.e., that each simplex can be involved in at most one pair) imply that there are significantly fewer valid local frame configurations than the possibilities provided by the bit flag representation. Thus, we can encode a local frame *compressed* representing only the valid configurations.

In 2D, for example, there are have 12 arrows for a total of $2^{12} = 4{,}096$ cases. However, for a Forman gradient, there are only 97 valid cases for a triangle. Thus, all possible configurations can be encoded by using only one byte per triangle. Similarly, in the 3D case, there are 32 arrows for a total of $2^{32} = 4{,}294{,}967{,}296$ possible configurations. Considering the valid configurations, there are only $51{,}030$ cases that can be represented with 2 bytes per tetrahedron.

5.3 Computing the Forman Gradient

We review the most relevant algorithms in the literature for computing discrete Morse and Morse-Smale complexes based on a Forman gradient. In order to apply Forman theory to a function f, which is given on the vertices (0-cells) of a simplicial complex Σ, the first step is to extend f to a function F defined on all cells of Σ (or to a corresponding Forman gradient vector field), such that $F(p) = f(p)$ for all vertices p of Σ. Note that such function F is not unique.

The algorithm described in Sect. 5.3.1 is defined for triangle meshes where the scalar function is computed as a pre-processing step. The algorithm presented in Sect. 5.3.2 has been defined in a dimension-independent way for d-dimensional simplicial complexes and computes the Forman gradient V based on the scalar values given at the vertices of the complex. The algorithms presented in Sect. 5.3.3 and Sect. 5.3.4, have been defined for computing a Forman gradient on cell complexes and implemented for regular grids. Both of them can be extended to the simplicial case.

5.3.1   Forman Approach Based on Connolly's Function

Forman theory has been used to build approximations of Morse and Morse-Smale complexes by Cazals et al. [2], with the objective of segmenting surfaces of 3D shapes representing molecules, where a scalar field, the Connolly's function, is computed. A triangle mesh Σ embedded in 3D space is decomposed into an approximation of a Morse-Smale complex using a discrete Connolly function f [3] computed at each vertex of Σ, which is then extended to the edges and triangles to a Forman function F.

There are two kinds of discrete gradient paths for F. The first kind contains vertices and edges, and the second kind contains triangles and edges. Thus, primal and dual graphs G and G^* of Σ are considered. A Forman gradient vector field on a closed 2-manifold is equivalent to a pair of interlaced primal and dual rooted forests, which are spanning forests of the primal and dual graph of Σ, respectively. There is one tree component T in the primal (dual) forest for each minimum (maximum) p of F, where p is the root of T. The discrete gradient vector field is directed from the leaves to the root in the primal forest, and from the root to the leaves in the dual forest. An example of interlaced forests of a Forman function F defined on a cube is illustrated in Fig. 5.4. Function F has one maximum (top face), one minimum (lower left vertex), and no saddles.

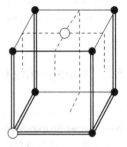

Fig. 5.4 Interlaced primal and dual forest of a Forman function F defined on a cube. All edges belong either to the primal forest (*double lines*), or to the dual forest (*dashed*). There is one critical vertex (*lower left vertex*), and one critical face (*top face*), both marked with a *hollow circle*

The algorithm, which computes a discrete gradient vector field on Σ, works in several steps. First, minima and maxima of F are found. Minima are found among the vertices of Σ (and of G), while maxima are found among the triangles of Σ

(vertices of G^*). Second, the dual spanning forest is built, with roots located at the maxima of F. Edges of Σ are processed in increasing order with respect to the function value, and are marked as visited and added to the dual forest if they do not connect components corresponding to different maxima of F. Third, the primal forest is constructed in a similar way, with minima of F as roots, considering only the edges not marked as visited.

Once a discrete gradient field of F is constructed, it can be used to define an approximation of the Morse-Smale complex on Σ. Since minima and maxima of F are located on vertices and faces of Σ, respectively, ascending and descending regions are defined on different elements of Σ. The ascending region of a minimum p is its component in the primal forest, and it consists of vertices and edges. The descending region of a maximum q is its component in the dual forest, and it consists of triangles and edges. These regions are not manifolds. Thus, each descending region is extended to a manifold by adding inner vertices and edges. Each ascending region is extended to a manifold by adding the triangles having all three vertices in the region. The boundary of each ascending cell is then composed of triangles. Each such triangle may remain unclassified (not belonging to any ascending cell), may be assigned to all ascending cells it bounds, or may be assigned to an ascending cell by a majority rule. An approximation of the Morse-Smale complex of f is defined as the intersection of ascending and descending manifolds.

5.3.2 A Forman-Based Approach for Tetrahedral Meshes

The algorithm proposed by King et al. in [11] takes as input a scalar field f defined over the vertices of a tetrahedral mesh Σ. It computes a Forman gradient V by subdividing the simplices of Σ into three lists, denoted as A, B and C, such that:

- lists A and B have the same length,
- for each i-simplex $\sigma_j \in A$, $(\sigma_j, \tau_j) \in V$, where τ_j is an $(i+1)$-simplex in B,
- C is the set of critical simplices.

The algorithm builds the Forman gradient V recursively on the *lower link* $Lk^-(v)$ of each vertex v in Σ (see Sect. 2.2.1). Lists A, B and C are initialized as empty. For each vertex v in Σ, if $Lk^-(v)$ is empty, then v is a minimum and it is added to C. Otherwise, v is added to A and the algorithm is recursively applied on the lower link $L^-(v)$, producing lists A', B', C'. Such lists define the Forman gradient V' on $Lk^-(v)$. The recursive call is performed until all the lower links are empty.

The Forman gradient V on v is computed as follows (the operation denoted by symbol $*$ takes two simplices and returns the simplex generated by the union of their vertices):

- the lowest critical vertex w is chosen from C' and the edge $v * w$ is added to B,
- for each i-simplex σ (different from w) in C', the $(i+1)$-simplex $v * \sigma$ is added to C,

- for each i-simplex σ in A' the $(i+1)$-simplex $v * \sigma$ is added to A and the $(i+2)$-simplex $v * V'(\sigma)$ is added to B.

The computed Forman gradient can produce an arbitrary large number of critical simplices compared to actual critical points of the original scalar function f. Thus, the algorithm uses a persistence value $\rho \geq 0$ to cancel critical simplices whose importance is too low. Intuitively, persistence measures the importance of a pair of critical cells (a definition of persistence is provided in the context of multi-resolution, in Sect. 6.1.3). Once the lower link of a vertex has been processed, a *persistence-based cancelling step* is performed. For each critical i-simplex σ, all the gradient paths to critical $(i-1)$-simplices are found. A critical i-simplex σ can be canceled with critical $(i-1)$-simplex γ if and only if there is only one gradient path from σ to γ. The effect of a cancellation is to reverse the gradient path connecting σ and γ. This is a discrete version of the *cancellation* operator described for Morse-Smale complexes in Sect. 6.1.1.

In Fig. 5.5, an example of the Forman gradient extraction on the link of a vertex is illustrated. The star of vertex 9 is shown in Fig. 5.5a. The application of the algorithm to the lower link of vertex 9, illustrated in Fig. 5.5b, produces the Forman gradient V' defined by the following lists :

$A' = 3; 4; 6; 7; 8$
$B' = [3,2]; [4,1]; [6,5]; [7,1]; [8,2]$
$C' = 1; 2; [4,3]; 5; [7,6]; [8,5]$

As shown in Fig. 5.5c, the lists are updated after the cancellations performed on V'. Vertex 2 is cancelled with edge $(3,4)$ while vertex 5 is cancelled with edge $(6,7)$. Then, V' is extended to V obtaining the following lists (shown in Fig. 5.5d).

$A = 1; [3,9]; [8,9]; [5,9]; [7,9]; [9,4]$
$B = [1,9]; [3,9,2]; [9,8,2]; [5,9,6]; [9,7,1]; [9,4,1]$
$C = [2,9]; [5,9]; [3,4,9]; [5,8,9]; [6,7,9]$

In Fig. 5.5e, the cancellation is performed also on V on the first pair of critical simplices, the edge $(2,9)$ and the triangle $(3,4,9)$. In Fig. 5.5f, the final gradient obtained after performing all the possible cancellations is shown.

5.3.3 The Algorithm by Gyulassy et al.

The algorithm presented by Gyulassy et al. in [7] computes the Morse-Smale complex starting from a regular d-dimensional cell complex Γ with scalar field f defined at the vertices of Γ. Function f is extended to a Forman function F, defined on all cells of Γ, such that $F(\sigma)$ is slightly larger than $F(\tau)$ for each cell σ and each face τ of σ. For such function F, all cells of Γ are critical. A discrete gradient vector field is computed by assigning gradient arrows in a greedy manner during sweeps over the cells of Γ according to increasing dimension and increasing F value. Each current non-paired and non-critical cell in the sweep is paired with

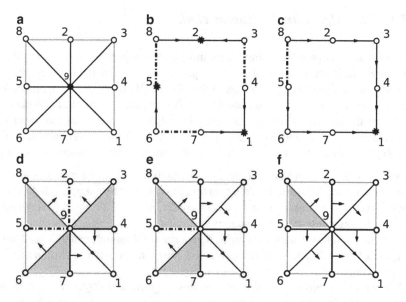

Fig. 5.5 (**a**) The lower star of vertex 9. The Forman gradient V' on the link of 9 before (**b**) and after (**c**) the cancellation of critical edge (4,3) and vertex 2, and edge (7,6) and vertex 5. The Forman gradient V (**d**) before the cancellation step, (**e**) after one cancellation and (**f**) the final Forman gradient V obtained in the lower star of vertex 9. The critical vertices and edges are depicted with *dotted lines* and critical triangles are depicted in *grey*

its co-face with only one face not marked (as critical or as already paired). If there are several of such co-faces the lowest one is taken. If there is no such co-face, a cell cannot be paired, and it is critical. Pairs built in this way define a discrete gradient vector field.

The order in which the cells in Γ are processed by the algorithm is not completely deterministic, since there could be many different i-cells in Γ with the same value of function F. As a consequence, some unnecessary critical cells may be produced by the algorithm. In [15, 16] a similar approach for computing a Forman gradient on two-dimensional and three-dimensional regular grids, respectively, has been defined dealing with this problem. A *weighted discrete function* is defined recursively on all the cells of the complex such that, when two cells share a common face whose function value is the maximum among both face sets, then the tie is broken using the second maximum face whose vertex sets are disjoint from the above common face. It has been proven that the pairs found by the algorithm are unique and independent of the order in which the cells are considered, thus providing a basis for parallelizing the algorithm.

5.3.4   The Algorithm by Robins et al.

Robins et al. [14] propose a dimension-independent algorithm for constructing a Forman gradient vector field on a regular grid K with scalar field values given at the vertices, and present applications to the 2D and 3D images. The approach has been extended to triangle and tetrahedral meshes in [10, 17]. We describe here the version of the algorithm for regular grids. The algorithm processes the lower star of each vertex v in K independently. For each cell σ in the lower star, the value $\max_{p \in \sigma} f(p) = fmax(\sigma)$ is considered. An ascending order is generated based on the values $fmax(\sigma)$ and the dimension of σ, such that each cell σ comes after its faces in the order. An example of such an order is obtained by listing the field values of the vertices of a simplex σ in increasing order, and considering a lexicographic order on the resulting strings of integers.

If the lower star of vertex v is empty, then v is a local minimum and it is added to the set C of critical cells. Otherwise, the first edge e in the order is chosen and pair (v, e) is added to V.

The star of v is processed using two queues, $PQone$ and $PQzero$, corresponding to i-cells with one and zero unpaired faces, respectively. All edges in the star of v different from e are added to $PQzero$. All co-faces of e are added to $PQone$ if the number of unpaired faces is equal to one. If queue $PQone$ is not empty, the first cell α is removed from the queue. If the number of unpaired faces of α has become zero, α is added to $PQzero$. Otherwise, $(\alpha, pair(\alpha))$ is added to V, where $pair(\alpha)$ is the unique unpaired face of α, $pair(\alpha)$ is removed from $PQzero$ and all the co-faces of either α or $pair(\alpha)$ with number of unpaired faces equal to one are added to $PQone$.

If $PQone$ is empty and $PQzero$ is not empty, one cell β is taken from $PQzero$. Cell β is added to the set C of critical points and all the co-faces of β with number of unpaired faces equal to one are added to $PQone$.

If both $PQzero$ and $PQone$ are empty, then the next vertex is processed. Result of the algorithm is the set C of critical cells and the pairing of non-critical cells, which define the Forman gradient vector field V.

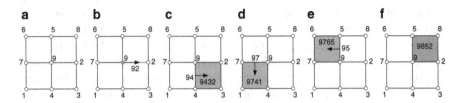

Fig. 5.6 Processing the lower star of vertex 9 using the algorithm in [14] (**a**) Input grid with values of f. (**b**) Vertex 9 is paired with edge 92. (**c**) Edge 94 is paired with triangle 9432. (**d**) Edge 97 is paired with triangle 9741. (**e**) Edge 95 is paired with triangle 9765. (**f**) Face 9852 is recognized as critical

In Fig. 5.6, we show the main steps of the algorithm in [14] when processing the lower star of vertex 9 (see Fig. 5.6a). Each vertex is labeled with its scalar field value. Other cells are labeled by the lexicographic order. The lower star of 9 is not empty, and thus 9 is not a minimum. The lowest edge starting from 9 (edge 92), is chosen to be paired with 9. All the other edges are inserted in $PQzero$ and the co-faces of 92 with a single unpaired face (faces 9432 and 9852) are inserted in $PQone$ (Fig. 5.6b). The first face is taken from $PQone$ (face 9432) and coupled with its single unpaired face (edge 94). The face 9741, which is a co-face of 94 with exactly one unpaired face, is inserted in $PQone$ and edge 94 is removed from $PQzero$ (Fig. 5.6c). Face 9741 is taken from $PQone$ and paired with edge 97, which is removed from $PQzero$. Face 9765 is inserted in $PQone$ and successively removed to be paired with edge 95 (Fig. 5.6d and e). Face 9852 is removed from $PQone$ and declared critical, as it has no unpaired faces (Fig. 5.6f).

In the 3D case, the algorithm in [14] does not create spurious critical cells. The extracted critical cells are in a one-to-one correspondence with the changes in topology in the lower level cuts of cubical complex K.

Weiss et al. [17] and Iuricich [10] have provided the first efficient implementation on tetrahedral meshes of the algorithm for computation of the combinatorial gradient vector field, based on the primal-dual data structure for representing discrete Morse complexes, described in Sect. 5.2.2.

5.4 Computing Discrete Morse and Morse-Smale Complexes

An additional advantage of Forman theory, in comparison with other approaches, is an efficient and elegant mechanism for retrieving the cells of the Morse complexes (i.e., the descending and ascending manifolds), the cells of the Morse-Smale complex, and the Morse incidence graph (see Sect. 1.3) from a d-dimensional complex Γ endowed with a local discrete gradient field encoded with a *compact gradient* [10, 17]. We discuss here how to extract the descending and ascending manifolds based on the cells of the primal complex Γ and the topological relations involved. Generally speaking, a descending or ascending i-cell is extracted by traversing the primal/dual complex following the pairs of the gradient field, and starting from the cell corresponding to the critical point associated with the descending/ascending i-cell. In 2D, all the V-paths can be visited in linear time visiting all the cell pairs at most once. However, in three dimensions and higher, gradient paths can branch and merge multiple times resulting in a many-to-many adjacency relationship, between critical cells not belonging to the extremal graph, leading to an overall extraction complexity of $O(m^2)$, where m denotes the number of vertices of Γ. For these reasons some proposal have been made for reducing the worst-case time complexity in 3D [5, 10, 15, 17].

5.4.1 Descending Morse Complex

The i-cells of the descending Morse complex Γ_d are naturally defined as a collection of primal i-cells of Γ. The computation of a descending i-cell always starts from a critical i-cell γ of Γ. All the $(i - 1)$-cells in the boundary of γ are selected and, among them, only the $(i - 1)$-cells paired with an i-cell different from γ are considered. From such i-cells, the breadth-first traversal of the complex continues until all the V-paths starting from γ have been visited.

Figure 5.7 shows the computation of a descending 2-cell on a triangle mesh Σ. The computation starts from a critical triangle (maximum) σ (dark triangle in Fig. 5.7). Among the edges on the boundary of σ, the edges paired with a triangle different from σ are considered. While navigating such arrows, the triangles reached are enqueued in a breadth-first traversal of the triangles of Σ until all the V-paths starting from σ have been visited.

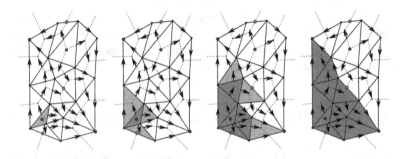

Fig. 5.7 Extraction of the descending 2-cell corresponding to a maximum (critical triangle)

5.4.2 Ascending Morse Complex

The i-cells of the ascending Morse complex Γ_a are naturally defined as a collection of i-cells of the dual complex Γ^* or equivalently as a collection of $(d - i)$-cells of Γ. The computation of an ascending i-cell starts from a critical $(d - i)$-cell γ of Γ. All the $(i + 1)$-cells in its star are selected and, among them, only the $(i + 1)$-cells paired with an i-cell different from γ are considered. From such i-cells the breadth-first traversal of the complex continues until all the V-paths ending in γ have been visited in reverse order.

Figure 5.8 shows the computation of an ascending 2-cell on a triangle mesh Σ. The computation starts from a critical vertex (minimum) σ (marked by a black dot in Fig. 5.8). All the edges on the immediate co-boundary of σ are considered; among them only the edges paired with a vertex different from σ are considered. Navigating such arrows, the vertices reached are enqueued in a breadth-first traversal of the vertices of Σ until all the V-paths ending in σ have been visited in reverse order.

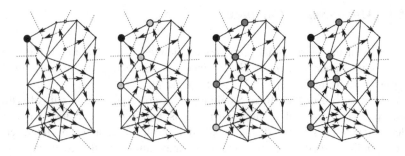

Fig. 5.8 Extraction of the ascending 2-cell corresponding to a minimum (critical vertex)

The computation of the ascending/descending Morse complex is performed through constant time operations on each cell on the V-paths visited. In 2D, the extraction of a descending i-cell is performed in linear time in the number of the i-cells, since they are visited only once traversing the V-paths. The complexity for the extraction of an ascending i-cell is the same.

In higher dimensions, the situation is more complex. For instance, in 3D, gradient paths can branch and merge, potentially resulting in many-to-many adjacency relationships between critical 1-cells and critical 2-cells. For example, let us consider a tetrahedral mesh Σ with m vertices whose discrete Morse function contains $O(m)$ critical 1-cells, each of which is connected to $O(m)$ critical 2-cells. This produces a discrete Morse complex containing $O(m^2)$ gradient paths between critical 1- and 2-cells. Since the number of critical 1- and 2-cells is bounded by m, the number of cells visited during the breadth-first search is also bounded by m and so the complexity of the whole extraction is in $O(m^3)$. Using a bit flag array to maintain the visited cells the standard breadth first traversal of the 1- and 2-cells can be employed. In [5, 17] the worst-case time complexity is reduced to $O(m_2)$, where m_2 denotes the number of 2-cells in Γ, forcing a single path not to visit the same cell more than once. Cells are marked, as visited or not visited by a V-path, and this leads to an increase in the storage. Overcoming this increase in the storage cost is crucial for a parallel implementation. The idea described in [16] is to interpret the substructure of the gradient paths as a Directed Acyclic Graph (DAG). Then, avoiding the standard breadth first traversal of the DAG, the path originating from a DAG node is visited only when all the paths entering it have been visited. In this way the common paths are visited only once and only the initial cells, where the common path starts, are visited more than once. This leads to a worst-case time complexity of $O(m_2 \log m_2)$ without increasing the storage cost.

5.4.3   Morse-Smale Complex

The i-cells of the Morse-Smale complex are defined as a collection of the i-cells of the *dually subdivided mesh* Γ_S obtained by intersecting the primal complex *Gamma* with its dual complex Γ^* (see Sect. 5.1.2).

A maximal cell of the Morse-Smale complex corresponds to a pair of critical points (a maximum and a minimum) and is encoded as a collection of so-called *micro-d-cells* in the dually subdivided mesh Γ_S obtained by intersecting the descending d-cell (corresponding to the maximum) and the ascending d-cell (corresponding to the minimum), which are collections of d-cells and vertices, respectively.

Let us consider a tetrahedral mesh. A 3-cell of the Morse-Smale complex corresponds to a pair of critical points (a maximum and a minimum) and is encoded as a collection of *micro-hexahedra* in the dually subdivided mesh Σ_S obtained by intersecting the descending 3-cell (corresponding to the maximum) and the ascending 3-cell (corresponding to the minimum), which are collections of 3-cells and vertices, respectively. Moreover, in the 3D case, the 3-cells of the Morse-Smale complex are bounded by a set of 2-cells corresponding to pairs of saddles (1-saddle and 2-saddle) and composed by a collection of *micro-quads*. Considering the primal/dual representation described in Sect. 5.1, for each pair of face-adjacent micro-hexahedra the common micro-quad is part of the Morse-Smale 2-cell if the labels of the two hexahedra are different.

The 1-skeleton of the Morse-Smale complex is composed of different types of 1-cells. For the 2D case, the 1-cells corresponding to a maximum-saddle or a minimum-saddle are the 1-manifolds of the ascending and descending Morse complex, respectively. In the 3D case, beside 1-cells corresponding to a maximum-2-saddle or a minimum-1-saddle, there are 1-cells called saddle connectors, which connect 1-saddles with 2-saddles. A saddle connector, between a 1-saddle p and a 2-saddle q, is computed by extracting the descending and ascending 2-manifolds associated with p and q. The descending 2-manifold extraction is performed first, and all the 2-cells traversed are marked as visited. Then, starting from the critical primal edge e corresponding to p and its adjacent edges, the same process as for extracting ascending 2-manifolds is performed, but only the 2-cells previously marked as visited are considered.

5.4.4   Morse Incidence Graph

The Morse Incidence Graph (MIG) represents the incidence relations between the cells of the Morse and Morse-Smale complexes defined on Γ (see Sect. 1.3). In the discrete case, such relations are computed traversing the V-paths of the *compact gradient V* defined on Γ, computing all the Morse cells in one of the two complexes, saving one node for each critical simplex and connecting two nodes in the graph with an arc if there is a separatrix V-path in V connecting the two corresponding critical simplices.

Let us consider the 2D case. When computing the MIG $G = (N, A, \varphi)$ on a two-dimensional scalar field on which a Forman gradient V has been defined, we start adding to N one node p for each critical cell γ in V. We consider the vertex v with highest function value in γ, and the index of v is stored in p. Then, for each maximum node p corresponding to a critical 2-cell γ, the descending 2-cell of γ is computed. The set of 2-cells visited during the V-path traversal are stored in the node and p is connected, with an arc in A, to all the saddle nodes corresponding to critical edges reached by V-paths. Dually, for each minimum node p corresponding to a critical vertex v, the ascending 2-cell of v is computed. The set of vertices visited during the V-paths traversal are stored in the node and p is connected, with an arc in A, to all the saddle nodes corresponding to critical edges reached by the V-paths ending at v. Note that the descending 2-cells are stored as collections of triangles while the ascending 2-cells are stored as collections of vertices (and, thus, of dual 2-cells).

In the 3D case, let us consider the *extremal graphs*, i.e., the subgraphs of the incidence graph defined between nodes corresponding to minima and those nodes corresponding to 1-saddles (or maxima and 2-saddles). The extremal graphs are dimension independent. In the 3D case, these two subgraphs of the MIG are computed in the same way as in 2D case. A new step is introduced to compute the saddle connectors, i.e., arcs of the MIG between 1-saddles and 2-saddles. Considering the 1-skeleton extraction of the Morse-Smale complex described in the previous section, saddle connectors are extracted in a similar fashion.

Figure 5.9 illustrates examples of the features extracted from the **Fighter** dataset, including the 3-cells of the Morse complex (Fig. 5.9b) and the combinatorial representation (Fig. 5.9c) of the MS complex.

Fig. 5.9 Example of features extracted from the **Fighter** dataset, representing data from a wind tunnel model developed at NASA Langley Research Center. (**a**) The original field values. (**b**) The Morse 3-cells, thresholded by region sizes to highlight the larger 3-cells decomposing the parts at higher turbulence. (**c**) The graph representing the combinatorial structure of the MS complex

5.5 Summary and Comparisons

In this chapter, we have described some key results obtained in the literature based on Forman theory. In Sect. 5.2, we have seen that a Forman gradient can be efficiently encoded for simplicial complexes as well as regular grids. The gap between the storage costs required for encoding a Forman gradient through the IA

Table 5.1 Evaluation of the data structures exemplified on some real datasets

Dataset	$|\Sigma_0|$	$|\Sigma_1|$	$|\Sigma_2|$	$|\Sigma_3|$	IA	IG
VisMale	5M	36M	62M	31M	1.09 GB	5.89 GB
Foot	5M	33M	57M	28M	1.01 GB	5.4 GB
Bonsai	5.9M	40M	69M	34M	1.21 GB	6.54 GB

For each dataset, we show number of vertices Σ_0, number of edges Σ_1, faces Σ_2 and tetrahedra Σ_3. Cost in Gigabytes (GB) of the IA and IG encoding the tetrahedral mesh Σ and the Forman gradient

Table 5.2 Algorithms for the extraction of Morse complexes and Morse-Smale complexes

Algorithm	Time	Input	Output
Cazals et al. [2]	$O(m \log m)$	Triangle mesh	Morse and Morse-Smale complexes
King et al. [11]	$O(m)$	Simplicial complex	Forman gradient
Gyulassy et al. [7]	$O(m)$	Cubic grid	Forman gradient
Robins et al. [14]	$O(m)$	Cubic grid	Forman gradient
Günther et al. [5]	$O(m_i)$	Cubic grid	Morse and Morse-Smale complexes
Shivashankar et al. [16]	$O(m_i \log m_i)$	Cubic grid	Morse and Morse-Smale complexes
Weiss et al. [17]	$O(m)$	Simplicial complex	Forman gradient

The reported time complexity is in the worst case, in terms either of the number m of vertices, or of the number m_i of i-cells of the input complex

data structure (or any other data structure which encodes only d-cells and 0-cells) and the Incidence Graph (IG), has been evaluated for some real datasets in [10]. Results are subsumed in Table 5.1. We can notice that the IG occupies four to five times more space than the IA data structure.

In Sect. 5.3, we have described the most widely used algorithms, based on Forman theory, to extract the cells in the Morse or Morse-Smale complexes, by computing first a Forman gradient V extending the scalar function f, defined on the vertices of a d-complex Γ, to all the cells of Γ.

Generally speaking, a scalar function defined on the vertices of a cell complex Γ is not required in order to compute a discrete Morse complex on Γ. This is the case where a discrete Morse complex is used for homology or persistent homology computation [8, 9]. Since we are interested in the analysis of a scalar field f, we focused on methods for computing a Forman gradient on the underlying geometry of the scalar field [2, 7, 11, 14]. Since Forman theory has been defined for cell complexes, most of the algorithms assume to work on generic cell complexes, although in practice they will be applied to regular grids or simplicial complexes.

The algorithms described in [7, 11, 14] process each vertex of Γ independently, resulting in a worst-time complexity of $O(m)$, with m number of vertices of Γ. The algorithm in [2], instead, needs an ordering of all the edges of the simplicial complex. It computes directly the ascending and descending regions, intersecting them to obtain the Morse-Smale 2-cells in a second step. Algorithms focused on the

Forman gradient computation need traversing the gradient paths to obtain the Morse and Morse-Smale cells.

In Table 5.2 we show a summary of the algorithms described in this chapter indicating the input format, the cells of the Morse or Morse-Smale complex computed as output and the worst-case time complexity. We denote with m the number of vertices of the d-dimensional complex Γ and with m_i the number of i-cells.

References

1. D. Canino, L. De Floriani, and K. Weiss. IA*: An adjacency-based representation for non-manifold simplicial shapes in arbitrary dimensions. *Computers & Graphics*, 35(3):747–753, 2011.
2. F. Cazals, F. Chazal, and T. Lewiner. Molecular shape analysis based upon the Morse-Smale complex and the Connolly function. In *Proc. 9th Annual Symposium on Computational Geometry*, pages 351–360, New York, USA, 2003. ACM Press.
3. M. L. Connolly. Measurement of protein surface shape by solid angles. *J. of Molecular Graphics*, 4(1):3–6, 1986.
4. H. Edelsbrunner. *Algorithms in Combinatorial Geometry*. Springer Verlag, Berlin, 1987.
5. D. Günther, J. Reininghaus, H. Wagner, and I. Hotz. Efficient computation of 3D Morse-Smale complexes and persistent homology using discrete Morse theory. *The Visual Computer*, 28(10):959–969, 2012.
6. T. Gurung and J. Rossignac. SOT: Compact representation for triangle and tetrahedral meshes. Technical Report GT-IC-10-01, College of Computing, Georgia Institute of Technology, Atlanta, GA, USA, 2010.
7. A. Gyulassy, P.-T. Bremer, B. Hamann, and V. Pascucci. A practical approach to Morse-Smale complex computation: Scalability and generality. *IEEE Transactions on Visualization and Computer Graphics*, 14(6):1619–1626, Nov-Dec 2008.
8. S. Harker, K. Mischaikow, M. Mrozek, and V. Nanda. Discrete Morse theoretic algorithms for computing homology of complexes and maps. *Foundations of Computational Mathematics*, 14(1):151–184, 2014.
9. S. Harker, K. Mischaikow, M. Mrozek, V. Nanda, H. Wagner, M. Juda, and P. Dlotko. The efficiency of a homology algorithm based on discrete Morse theory and coreductions. In *Proc. 3rd International Workshop on Computational Topology in Image Context (CTIC 2010). Image A*, volume 1, pages 41–47, 2010.
10. F. Iuricich. *Multi-resolution shape analysis based on discrete Morse decompositions*. PhD thesis, University of Genova – DIBRIS, Italy, 2014.
11. H. King, K. Knudson, and N. Mramor. Generating discrete Morse functions from point data. *Experimental Mathematics*, 14(4):435–444, 2005.
12. G. M. Nielson. Tools for triangulations and tetrahedralizations and constructing functions defined over them. In G. M. Nielson, H. Hagen, and H. Müller, editors, *Scientific Visualization: overviews, Methodologies and Techniques*, chapter 20, pages 429–525. IEEE Computer Society, Silver Spring, MD, 1997.
13. A. Paoluzzi, F. Bernardini, C. Cattani, and V. Ferrucci. Dimension-independent modeling with simplicial complexes. *ACM Transactions on Graphics*, 12(1):56–102, January 1993.
14. V. Robins, P. J. Wood, and A. P. Sheppard. Theory and algorithms for constructing discrete Morse complexes from grayscale digital images. *IEEE Transactions on Pattern Analysis and Machine Intelligence*, 33(8):1646–1658, 2011.

15. N. Shivashankar, S. Maadasamy, and V. Natarajan. Parallel computation of 2D Morse-Smale complexes. *IEEE Transactions on Visualization and Computer Graphics*, 18(10):1757–1770, 2012.
16. N. Shivashankar and V. Natarajan. Parallel computation of 3D Morse-Smale complexes. *Computer Graphics Forum*, 31(3):965–974, 2012.
17. K. Weiss, F. Iuricich, R. Fellegara, and L. De Floriani. A primal/dual representation for discrete Morse complexes on tetrahedral meshes. *Computer Graphics Forum*, 32(3):361–370, 2013.

Chapter 6
Simplification and Multi-Resolution Representations

Although Morse and Morse-Smale complexes encode the behavior of a scalar field f and the topology of its domain while abstracting from fine geometric details, simplification of these complexes is an important issue to eliminate noise and reduce over-segmentation. Moreover, different users usually have different requirements in terms of degree of simplification, which usually vary over time and location within the field domain. Thus, a multi-resolution representation of the morphology of the field is critical for supporting interactive analysis and exploration of data.

In Sect. 6.1, we describe two simplification operators for Morse functions, and for Morse and Morse-Smale complexes, called *cancellation* and *remove*, respectively, and we present some results from their experimental comparison.

In Sect. 6.2, we present multi-resolution models for the morphology of scalar fields. These have been designed mainly for two-dimensional scalar fields (i.e., terrains). We describe two models in more detail: the first one provides a multi-resolution description of the Morse incidence graph in arbitrary dimensions, while the second one addresses the problem of coupling a multi-resolution representation of the geometry and of the morphology of a terrain.

6.1 Simplification Operators

We describe here two simplification operators for Morse and Morse-Smale complexes proposed in the literature, namely *cancellation* [18] and *remove* [6]. Both operators eliminate a pair of critical points of a scalar field, which corresponds to the elimination of a pair of cells in the Morse complexes and of a pair of vertices in the Morse-Smale complex. The difference is that *cancellation* often introduces a large number of cells (of dimension higher than zero) in the Morse-Smale complex,

© The Author(s) 2014
L. Čomić et al., *Morphological Modeling of Terrains and Volume Data*,
SpringerBriefs in Computer Science, DOI 10.1007/978-1-4939-2149-2_6

while this never happens with *remove*. On large data sets, *cancellation* can create complexes that exceed practical memory capabilities [17], while *remove* always reduces the size of the complexes generated by applying it.

6.1.1 cancellation Operator

The *cancellation* operator [20] simplifies a Morse function f defined on a manifold M by eliminating two critical points p and q and locally modifying the integral lines originating at or converging to them [20]. The two critical points p and q can be cancelled if

1. p is an i-saddle and q is an $(i + 1)$-saddle, and
2. p and q are connected through a unique integral line of f.

After the *cancellation* of p and q, each critical point r of index at least $i + 1$, which was connected through an integral line to i-saddle p, becomes connected to each critical point of index at most i, which was connected to $(i + 1)$-saddle q before the *cancellation*.

In Morse complexes, a *cancellation* removes two cells both in the ascending and descending Morse complexes, and locally modifies the incidence relation between the remaining cells. In a descending (and, symmetrically, ascending) Morse complex, an i-cell p and an $(i + 1)$-cell q can be cancelled if cell p appears exactly once on the boundary of cell q. After *cancellation*, each cell r, which was on the boundary of $(i + 1)$-cell q, becomes part of the boundary of each cell s which was in the co-boundary of i-cell p. In the Morse-Smale complex, there is a new k-cell for each two cells s and r that become incident to each other in the Morse complexes after *cancellation* and that differ in dimension by k.

In the Morse incidence graph G representing the incidence relation (encoded by arcs in A) between the cells in the Morse complexes (encoded by nodes in N), a *cancellation* deletes two nodes from N, and locally modifies the connection between the neighboring nodes. An i-node p and an $(i + 1)$-node q can be cancelled if node p is connected to node q through an arc with label equal to 1. After *cancellation*, each i-node r, which was connected to $(i + 1)$-node q, becomes connected to each $(i + 1)$-node s which was connected to i-node p.

In 2D, a *cancellation* removes an extremum and a saddle. A cancellation of a maximum p and a saddle q is *feasible* on a Morse incidence graph if 1-node q is connected with two different 2-nodes p and p'. After cancellation, 1-node q (saddle) and 2-node p (maximum) are deleted. The corresponding cells in the descending Morse complex are contracted into the cell represented by 2-node p'. All the 1-nodes connected to p before *cancellation* are redirected to p'. A *cancellation* of a minimum p and a saddle q is entirely dual.

In Fig. 6.1, we show an example of a cancellation of a maximum p and a saddle q. In the descending Morse complex, shown in Fig. 6.1 (left), 1-cell q is deleted and 2-cell p is merged with the other 2-cell p' incident in q. In the Morse incidence graph (see Fig. 6.1 (right)), nodes p and q are deleted as well as all their incident arcs. Nodes r_1, r_2 and r_3 are connected to p' as the corresponding 1-cells become part of the boundary of p'.

In 3D, there are two instances of *cancellation*: one instance applies to an extremum and a saddle (a maximum and a 2-saddle, or a minimum and a 1-saddle), the other instance applies to two saddle points. The first instance of *cancellation* is exactly the same as in 2D case.

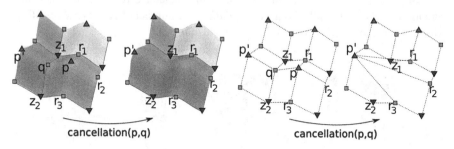

Fig. 6.1 Cancellation of a maximum p and a saddle q on a 2D descending Morse complex (*left*) and on the corresponding Morse incidence graph (*right*). Colored regions correspond to the 2-cells of the descending Morse complex Γ_d. Critical points are depicted as blue down-pointing triangles (minima), green squares (saddles) and red up-pointing triangles (maxima). (Color figure online)

The second instance of *cancellation* is more involved. The number of cells in the Morse complexes that become incident to each other (and thus, the number of cells in the Morse-Smale complex and the number of arcs in the Morse incidence graph) may increase after such a *cancellation*. Let p and q be a 1-node (1-saddle) and a 2-node (2-saddle), respectively. Let $R = \{r_j, j = 1, .., j_{max}\}$ be the set of 2-nodes connected to p and different from q, and let $T = \{t_k, k = 1, .., k_{max}\}$ be the set of 1-nodes connected to q and different from p. The application of *cancellation* to 1-node p and 2-node q removes both p and q from N and removes $j_{max} + k_{max} + 1$ arcs from A, but it adds $j_{max}k_{max}$ arcs to A, i.e., one for each pair (r_j, t_k), where $r_j \in R$ and $t_k \in T$.

In Fig. 6.2, we show an example of a *cancellation* of a 1-saddle p and a 2-saddle q. In the Morse incidence graph (see Fig. 6.2, left), nodes p and q are removed and the nodes r_1, r_2 and r_3, previously connected to p, become connected to nodes t_1, t_2 and t_3, previously connected to q. In the corresponding descending Morse complex (see Fig. 6.2, right) 1-cell p and 2-cell q are removed and the 2-cells r_1, r_2 and r_3 are stretched to fill the space of q, with t_1, t_2 and t_3 becoming part of their boundary.

Several strategies have been proposed in [17], which aim at postponing a *cancellation* that would introduce a number of arcs greater than a predefined threshold, or nodes having more than a certain number of incident arcs.

6.1.2   *remove Operator*

In [7], a new dimension-independent simplification operator, called *remove*, has been defined. It has two instances, namely $remove_{i,i+1}$ and $remove_{i,i-1}$, for $1 \leq i \leq d - 1$, defined on a d-dimensional Morse function f, on its associated Morse complexes and on the corresponding Morse incidence graph.

In Morse complexes, a *remove* deletes two cells both in ascending and descending Morse complexes, and locally modifies the incidence relations between the remaining cells. Differently from a *cancellation*, the number of cells in the resulting Morse-Smale complex is constantly reduced also in 3D.

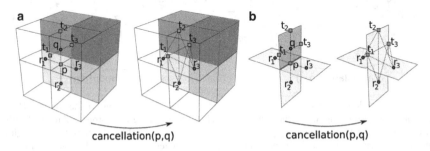

Fig. 6.2 Application of *cancellation* of a 1-saddle p and a 2-saddle q on a Morse incidence graph (**a**) and on the 1- and 2-cells of the descending Morse complex (**b**) in 3D. Colored cubes correspond to the 3-cells in the descending Morse complex Γ_d. Critical points are depicted as green squares (1-saddles) and purple dots (2-saddles). *Dotted lines* correspond to the arcs of the Morse incidence graph. (Color figure online)

On the descending Morse complex, $remove_{i,i+1}$ operator deletes an i-cell q and an $(i + 1)$-cell p, where q appears exactly once on the boundary of p. It is feasible if i-cell q is incident to at most one other $(i + 1)$-cell p' different from $(i + 1)$-cell p. It merges $(i + 1)$-cell p into $(i + 1)$-cell p' by deleting i-cell q. Operator $remove_{i,i-1}(q, p, p')$ is dual to the previous one.

On the Morse incidence graph, $remove_{i,i+1}(q, p, p')$ deletes an i-node q and an $(i + 1)$-node p if and only if q is connected to p through an arc labeled with 1, and to at most one $(i + 1)$-node p' different from p. We denote as R the set of i-nodes r_j connected with p, Z the set of $(i - 1)$-nodes z_k connected with q and S the set of $(i + 2)$-nodes s_h connected with p. Operator $remove_{i,i+1}(q, p, p')$ deletes all the arcs of the form (p, s_h), (q, z_k), (p, q) and (p', q). Each node r_j connected to p with an arc of the form (p, r_j), is redirected to p' with an arc (p', r_j).

In 2D, *remove* is equivalent to *cancellation* described in Sect. 6.1.1: $remove_{1,2}$ is equivalent to the cancellation of a saddle and a maximum, while $remove_{1,0}$ is equivalent to the cancellation of a saddle and a minimum.

In 3D there are four *remove* operators. $remove_{2,3}(q, p, p')$ and $remove_{1,0}$ (q, p, p') are the 3D extension of the same operators present in 2D. They are equivalent to the cancellation of a 2-saddle and a maximum, and to the cancellation of a 1-saddle and a minimum, respectively. Operators $remove_{1,2}(q, p, p')$ and $remove_{2,1}(q, p, p')$ are introduced specifically for the 3D case and remove a 1-saddle and a 2-saddle. On the descending complex, $remove_{1,2}(q, p, p')$ is feasible if 1-cell q is shared by exactly two different 2-cells p and p', and if q appears exactly once in the boundary of p. The operator removes 1-cell q and merges 2-cell p into p'. The boundary of p becomes part of the boundary of p'. Operator $remove_{2,1}(q, p, p')$ is dual.

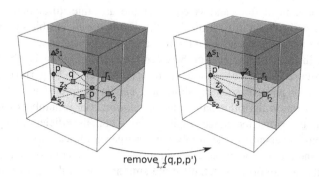

Fig. 6.3 $remove_{1,2}(q, p, p')$ of a 2-saddle p and a 1-saddle q in 3D. Colored regions correspond to 3-cells of the descending Morse complex Γ_d. Critical points are depicted as blue down-pointing triangles (minima), green squares (1-saddles), purple dots (2-saddles) and red up-pointing triangles (maxima). *Dotted lines* correspond to the arcs of the Morse incidence graph. (Color figure online)

In Fig. 6.3 we show an example of a $remove_{1,2}(q, p, p')$ in 3D. In the descending Morse complex, the 2-cells corresponding to p and p' are merged by deleting q. In the Morse incidence graph, nodes p and q are removed and all the arcs connecting them with the nodes in Z and S are removed as well. Nodes r_1, r_2 and r_3 are then connected with p'.

6.1.3   Comparison of cancellation and remove Operators

For three and higher dimensional scalar fields, Morse complexes admit a number of feasible *cancellation* operators greater than or equal to the number of feasible *remove* operators: a *cancellation* is feasible for any p and q that are connected through one integral line, while a *remove* is feasible if, in addition, q is connected to at most one saddle p' different from p of the same index as p. As a result, *remove* operator constantly reduces the number of arcs in the Morse incidence graph and, consequently the number of cells in the corresponding Morse-Smale complex. The same does not hold using *cancellation*, as shown in Sect. 6.1.1.

Two macro-operators have been defined in [19] to increase the number of viable *remove* simplifications at any time. These macro-operators are implemented as a sequence of extremum-saddle operators followed by a saddle-saddle operator. The first macro-operator collapses a 2-saddle p and a 1-saddle q into another 2-saddle p'. For all the 2-saddles t_j connected to q, a $remove_{2,3}(t_j, p_1, p'_1)$ is performed to eliminate the 2-saddles connected to q and different from p and p'. When p and p' are the only 2-saddles connected to q, the $remove_{1,2}(q, p, p')$ is performed. The second macro-operator is dual.

In [8], two simplification algorithms are developed, based on *remove* and *cancellation*, respectively. A *persistence* value is associated with each simplification operator. Persistence measures the importance of the pair of critical points, candidate for cancellation, and is equal to the absolute difference in function values between them. Simplification algorithms compute all feasible simplifications, evaluate their persistence and insert them in an ordered queue in increasing order of persistence. In [8, 19] a comparison between the two algorithms has been done. It has been observed that saddle-saddle operators are likely to be performed early in the simplification process (such simplifications can be interpreted as noise removal). With a simplification algorithm based on *cancellation*, a large number of arcs will be introduced in the Morse incidence graph, affecting both the efficiency (speed) of the algorithm and its versatility (the number of feasible simplifications). When the data set is small and the number of simplifications is high compared to the total number of nodes, the two algorithms behave quite similarly. With the growth of the size of the data set they start to differ: by using *remove* instead of *cancellation*, a 20 % more compressed Morse incidence graph can be obtained in about half the time. On several data sets, it has been noticed that by using *cancellation* the number of arcs remains approximately the same while by using *remove* their number immediately decreases.

6.2 Multi-Resolution Models

The first hierarchical representation of scalar fields, called hierarchical watershed representation [1], can be found in image analysis to reduce over-segmentation naturally present when watershed transformation is used as a segmentation tool. Most common hierarchical models for Morse complexes have been applied to terrain modeling. Such models can be applied to 2-manifolds with boundary embedded in 3D space and endowed with a scalar field [2, 4, 7, 11, 12, 14].

In Sect. 6.2.1, we review some approaches to the multi-resolution representations of 2D scalar fields proposed in the literature. In Sect. 6.2.2, we discuss a topological multi-scale model for scalar fields in arbitrary dimensions. In Sect. 6.2.3, we discuss a combined topological and geometric multi-scale model for terrains.

Before that, we introduce the basic idea underlying all such multi-resolution models. They all rely on a sequence of simplification operators applied to an initial representation at full resolution of the reference structure (e.g., the Morse-Smale complex, the Morse incidence graph, etc.), and producing a final coarse representation of the same structure. They consider:

- The coarse representation, called *base*;
- The reverse *refinement operators* of the performed simplification operators (corresponding to recovering the full resolution structure from the base one); they are also called *modifications*;
- A *direct dependency relation* among refinement operators, which is used to relax the total order of refinements into a partial order;
- A Directed Acyclic Graph (*DAG*) where the nodes are the refinement operators and the arcs encode the direct dependency relation: there is an arc (μ_i, μ_j) if and only if μ_j directly depends on μ_i. The DAG is rooted and its root is a dummy modification corresponding to the creation of the base.

A large number of adaptive representations can be extracted from the DAG by considering all the closed sets of DAG nodes (refinement operators), where a set U of DAG nodes is closed if, for $\mu_i \in U$, U contains all source nodes of arcs entering in μ_i.

A closed subset of refinements (DAG nodes) can be applied to the base in any total order that extends the partial order (encoded in DAG arcs), thus producing a representation at an intermediate resolution. This provides a collection of representations at intermediate *uniform* and *variable* levels of detail which can be obtained from DAG. A *selective refinement query* consists of extracting the least refined representation (among those which can be obtained from closed sets of the DAG), satisfying some application-dependent criterion. Such a query is answered through a DAG traversal process.

6.2.1   Models for 2D Scalar Fields

In [2, 3], a multi-resolution morphological model for 2D scalar fields has been described, which is based on the *cancellation* operator (described in Sect. 6.1.1) and on the Morse-Smale complex.

The basic component is a *diamond*, which consists of a saddle s, its adjacent minima and maxima, and the integral lines connecting them. The lines that connect the saddle s to the extremal points are the separatrix lines incident in s, while the lines that connect the extremal points are chosen arbitrarily among the integral lines in the corresponding quadrangular 2-cell (slope district) in the Morse-Smale complex. Thus, each quadrangular 2-cell in the Morse-Smale complex is split in two triangles, by an arbitrary integral line which connects the vertex of the 2-cell corresponding to a minimum to the vertex corresponding to a maximum, as illustrated in Fig. 6.4a. The four triangles incident in the same saddle are grouped together in a diamond. The possible types of diamonds for a Morse-Smale function f are illustrated in Fig. 6.4b.

The dependency relation between refinements is defined as follows: two refinements are dependent if the associated diamonds have at least one vertex in common [2], or if they have an edge in common [3]. The dependency relation in [3] is clearly less restrictive than the one in [2], and this affects the number of morphological representations of the scalar field encoded in the multi-resolution model.

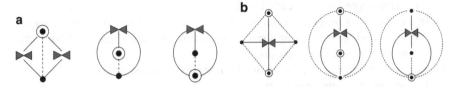

Fig. 6.4 (a) Splitting a quadrangular 2-cell in two triangles: a generic slope district; an isolated mountain; a crater. (b) The possible types of diamonds: all four 2-cells incident in the saddle in the Morse-Smale complex are generic; one of the 2-cells incident in the saddle is an isolated mountain; one of the 2-cells incident in the saddle is a crater

There have been some proposals in the literature to modify not only the Morse and Morse-Smale complexes using *cancellation*, but to modify also the scalar function f, thus constructing a new function g that corresponds to the simplified field. The first work in 2D, presented in [2] and improved in [21], modifies function f numerically, using Laplacian smoothing. In [2], function f is locally modified after each *cancellation* in order to agree with the new topology, by minimizing the error and obtaining a smooth approximation. In [21], the bottleneck of the smoothing step performed after each *cancellation* in [2] is solved by constructing a topologically valid function after all *cancellation* steps. The two resulting C^0 functions are comparable but the algorithm in [21] is faster. Moreover, in [21], a novel schema is devised to provide C^1-continuity.

Another approach, presented in [15] for the 2D case, modifies the scalar field f combinatorially, by changing the order in which the vertices appear in their sorted list according to function values.

6.2.2 A Multi-Resolution Morphological Model
for Arbitrary Scalar Fields

The *Multi-resolution Morse Incidence Graph* (*MMIG*) is a multi-resolution model representing the morphology of the Morse complexes as well as of the 1-skeleton of the Morse-Smale complex at different level of detail [7]. The *MMIG* is generated from the Morse Incidence Graph (*MIG*) representing the Morse complexes at full resolution by iteratively applying the simplification operator *remove* discussed in Sect. 6.1.2 (see Fig. 6.5). The *MIG* obtained as a result of the simplification sequence is the coarsest representation of the two complexes. It is denoted as $G_B = (N_B, A_B, \varphi_B)$, and is the *base* of the multi-resolution model.

The DAG nodes in the *MMIG* are the inverse refinements of the *remove* operators applied during simplification. The *dependency relation* is defined as follows. Let μ be a refinement modification, which introduces nodes p and q, inverse to a *remove*, and let $N^- = \{p'\} \cup R \cup S \cup Z$ (see Sect. 6.1.2). Refinement modification μ *directly depends* on refinement modification μ' if and only if μ' creates one or more nodes in N^-.

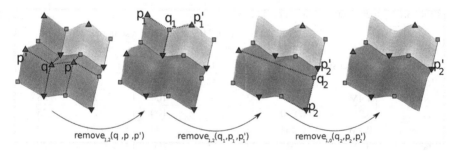

Fig. 6.5 Sequence of $remove_{1,2}(q, p, p')$, $remove_{1,2}(q_1, p_1, p'_1)$ and $remove_{1,0}(q_2, p_2, p'_2)$ performed on a relevant part of the Morse incidence graph encoding the descending Morse complex. For each remove operator, dotted lines connect p and q to the set of nodes considered for the dependency relation (set N^-)

It has been shown in [5] that the dependency relation defining the *MMIG* is less restrictive than the one in [2], while there is no containment with the dependency relation in [3].

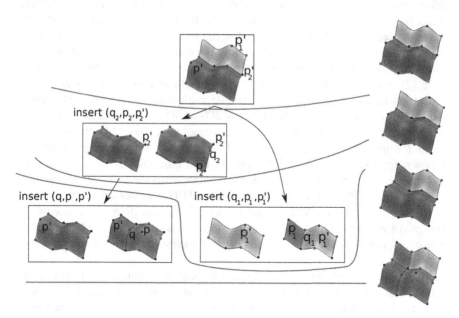

Fig. 6.6 *MMIG* structured as a Direct Acyclic Graph (DAG) where the root encodes the base complex, each node encodes a refinement modification and each directed arc indicates a dependency relation

Figure 6.6 shows an example of an *MMIG* built from the sequence of simplifications illustrated in Fig. 6.5. The refinement modifications are shown on the descending Morse complex for clarity. In the root, the base complex is stored. Each

Table 6.1 Experimental results from 2D data sets

| | | | Extracted *MIG* at different resolutions | | | | | |
| | | | Uniform | | | Variable | | |
Dataset name	Dataset size	*MMIG* size	0.1 (%)	0.001 (%)	0 (%)	0.1 (%)	0.001 (%)	0 (%)
Baia	1.4M	2, 410	27	42	100	12	17	27
Baia	4.4M	9, 333	22	36	100	18	25	38
Baia	8.3M	14, 940	20	31	100	18	21	35
Marcy	0.4M	2, 264	32	61	100	15	20	29
Marcy	0.8M	9, 012	29	63	100	16	19	28
Marcy	1.8M	11, 491	25	55	100	13	17	21
Maui	0.6M	8, 673	20	55	100	15	21	38
Maui	3.4M	16, 540	23	43	100	12	18	24
Maui	4M	16, 623	24	42	100	15	19	25

Sizes of data sets expressed as number of triangles, *MMIG* sizes expressed as number of DAG nodes, sizes (expressed as a percentage of the MIG at full resolution) of Morse incidence graphs extracted at the various persistence thresholds both at uniform and at variable resolution

refinement encoded in the model modifies only a local subset of the complex and two refinements are connected by an arrow if and only if the refinement pointed by the arrow depends from the other one. The red lines indicate all closed sets of the *MMIG* (each closed set contains the nodes above the line), and the corresponding complexes, extracted by performing the refinements of the set, are shown.

In [7], results are presented from the extraction of representations at various persistence levels at both uniform and variable resolution (with persistence threshold satisfied in 1/10 of the domain) from several 2D and 3D data sets. Variable resolution queries are the distinctive feature of a multi-resolution model like the *MMIG*. In Tables 6.1 and 6.2, the persistence threshold is indicated as a percentage of the field values range (0.1, 0.001, 0.0001, 0), and for each threshold, the *MIG* is extracted concentrating the resolution through the domain (column *uniform*) or inside a window query (column *variable*). The resulting *MIG* is expressed as a percentage of the size of the *MIG* at full resolution. A dramatic reduction of the size of the extracted *MIG* at relatively low persistence thresholds at uniform resolution can be noticed, especially for 2D data sets, indicating the presence of high level of noise in the data sets.

The results of applying the selective refinement algorithm on **Matterhorn** 2D data set are illustrated in Fig. 6.7. The descending Morse complex is shown, extracted after 10 refinements (Fig. 6.7a), 300 uniformly distributed refinements (Fig. 6.7b) and 1,200 refinements performed in a selected subset of the domain (Fig. 6.7c).

Table 6.2 Experimental results from 3D data sets

Dataset name	Dataset size	MMIG size	Extracted MIG at different resolutions					
			Uniform			Variable		
			0.1 (%)	0.01 (%)	0.0001 (%)	0.1 (%)	0.01 (%)	0.0001 (%)
VisMale	1.4M	5,271	12	52	94	9	21	55
VisMale	9M	20,559	13	48	87	8	21	34
VisMale	31M	32,612	32	62	92	18	22	31
XMasTree	2.7M	44,719	27	59	81	9	15	24
XMasTree	10M	121,193	18	37	78	19	18	24
XMasTree	24M	193,338	15	26	68	3	9	13
Bonsai	2.1M	29,127	22	43	79	12	29	38
Bonsai	20M	43,761	17	24	81	9	19	25
Bonsai	28M	45,761	11	26	81	8	12	24

Sizes of data sets expressed as number of tetrahedra, *MMIG* sizes expressed as number of DAG nodes, sizes (expressed as a percentage of the MIG at full resolution) of Morse incidence graphs extracted at the various persistence threshold both at uniform and at variable resolution

Fig. 6.7 (**a**) The descending Morse complexes extracted after 10 refinements, (**b**) after incrementing the resolution through the whole dataset, and (**c**) at variable resolution, for the **Matterhorn** terrain data set

6.2.3 A Combined Morphological and Geometrical Multi-Resolution Model for Triangulated Terrains

In the literature, the problem of simplification (and multi-resolution) has mainly been addressed for geometry and for morphology independently. The approach in [9] considers an initial triangle mesh Σ with associated Morse-Smale complex, and simplifies it through iterative vertex removal while considering Σ as a constrained triangulation with separatrix lines acting as constraints. Only the geometric structure of separatrix lines is simplified simultaneously with the triangle mesh, while the combinatorial structure of the Morse-Smale complex (described by the Morse incidence graph) remains the same across all levels of resolution.

A first attempt to couple the inspection of both the topology and the geometry of a terrain has been studied by Danovaro et al. in [10]. They proposed a model, called a *Multi-resolution Morse Triangulation (MMT)*, which manages geometry and morphology in a combined and consistent way, and, thus, is able to dynamically select consistent geometrical and morphological representations at a desired resolution. Such a model is presented for 2D scalar fields (terrains), but it can be generalized to higher dimensions as well.

The *MMT* [7, 19] combines the Multi-Triangulation (*MT*) proposed for triangle meshes in [13] and a 2D version of the Multi-resolution Morse Incidence Graph (*MMIG*).

The construction of an *MMT* starts from a triangle mesh Σ representing a 2D scalar field f, and its morphological model in the form of the 1-skeleton K of the Morse Smale complex (called *critical net*, see Sect. 1.3) plus the Morse Incidence Graph (*MIG*) G. We recall that the *MIG* is a purely combinatorial structure, while the critical net also includes the geometry of separatrix lines.

The three components of the model are simultaneously simplified in a combined and consistent way. The simultaneous simplification of Σ, K and G is driven by geometry. Triangle mesh Σ is simplified through a sequence of *half-edge collapse* operations, driven by quadric error metric for accurate error estimation [16]. A half-edge collapse, denoted $(v_i, v_j) \rightarrow v_j$, collapses two vertices v_i and v_j connected by an edge in Σ, into v_j. Triangles incident in edge (v_i, v_j) disappear, and triangles incident only in v_i become incident in v_j. The algorithm selects an edge to collapse on Σ according to the quadric error metric, and checks the feasibility of such simplification on K and G. The simplification proceeds as long as there are edges to be simplified. Simplification operations can be of three types:

1. operations which modify just Σ (they involve vertices lying in the interior of the ascending/descending 2-cells);
2. operations which modify Σ and K (they occur in the interior of the ascending/descending 1-cells);
3. operations which modify Σ, K and G (they involve critical vertices, i.e., ascending/descending 0-cells).

Simplifications of type 1 are simply half-edge collapses on Σ. Simplifications of type 2 modify triangle mesh Σ and the geometry of the separatrix lines in K. They consist of half-edge collapses on both Σ and K. Simplifications of type 3 collapse a critical point from the 1-skeleton of the Morse Smale complex and, thus, trigger a simplification operation on the *MIG*, collapsing a minimum and a saddle into an adjacent minimum, or a maximum and a saddle into an adjacent maximum.

The *MMT* multi-resolution model consists of three DAGs $\mathcal{M}_\Sigma$, $\mathcal{M}_K$ and $\mathcal{M}_G$, with mutual cross-links. The nodes of such DAGs are the inverse modifications of simplifications of all types for $\mathcal{M}_\Sigma$, of types 2 and 3 for $\mathcal{M}_K$, and of type 3 only for $\mathcal{M}_G$. They are connected through arcs which encode a dependency relation based on the vertices of Σ. Modifications of the three DAGs are linked together. Each node of $\mathcal{M}_K$ has a link to the node of $\mathcal{M}_\Sigma$ corresponding to the same operation performed during simplification. Similarly, each node of $\mathcal{M}_G$ has a link to the node of $\mathcal{M}_K$ corresponding to the same operation.

The purpose of combining geometry and morphology in a multi-resolution model is to support the extraction of a geometric representation that conforms to any simplified morphology. In other words, the level of detail on the morphological

representation is selected and then a simplified geometrical representation is
extracted which is consistent with the morphological one. Selective refinement
on the *MMT* produces a triangle mesh Σ', the 1-skeleton K' of its Morse-Smale
complex, and its Morse incidence graph G'.

While the construction is guided by geometric error, selective refinement on
the *MMT* is guided by morphological criteria. A persistence threshold is given for
the morphological component $\mathcal{M}_G$. This defines a closed set U_G of nodes of $\mathcal{M}_G$,
corresponding to applying all refinement modifications with lower persistence than
the threshold (plus other nodes needed to make a closed set). By exploiting cross
links connecting nodes of $\mathcal{M}_G$ to nodes of $\mathcal{M}_K$, U_G defines a closed set U_K of
nodes of $\mathcal{M}_K$, containing all nodes of $\mathcal{M}_K$ which are linked from nodes of U_G (plus
other nodes needed to make a closed set). In the same way, by exploiting cross links
from nodes of $\mathcal{M}_K$ to nodes of $\mathcal{M}_\Sigma$, set U_K induces a closed set U_Σ in $\mathcal{M}_\Sigma$. The
algorithm performing such synchronized traversal of the three DAGs is explained in
details in [10]. The results of the algorithm are: the morphological representation G'
of minimum size satisfying the persistence threshold, the critical net K' of minimum
size that is compatible with G' (i.e., where the path of integral lines is least refined),
and the triangle mesh Σ' of minimum size that is compatible with K (i.e., where
the surface inside Morse regions is the least detailed).

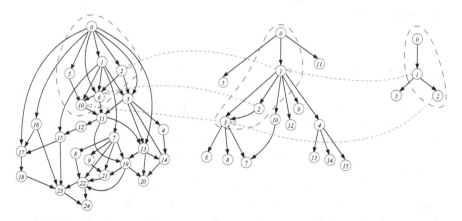

Fig. 6.8 Selective refinement on a *MMT*. The *red dotted line* encloses the closed sets of
modifications U_G extracted from $\mathcal{M}_G$, U_K induced by U_G on $\mathcal{M}_K$, and U_Σ induced by U_K on
$\mathcal{M}_\Sigma$. Only relevant cross links are shown. (Color figure online)

A working example is shown in Fig. 6.8: $\mathcal{M}_G$ is traversed and modifications 1 and
2 are added to the current closed set U_G. Since modifications 1 and 2 are associated
with modifications 1 and 3 of $\mathcal{M}_K$, 1 and 3 are added to U_K, along with their
ancestors 0 and 2. Modifications 1,2, 3 of $\mathcal{M}_K$ are associated with modifications
2,6,10 of $\mathcal{M}_\Sigma$, so such modifications, and their ancestors, are added to U_Σ. Results
shown in [10] on terrains report that, in terms of number of DAG nodes, $\mathcal{M}_G$ is

from 2 % to 9 % of $\mathscr{M}_K$, and from 0.2 % to 3 % of $\mathscr{M}_\Sigma$; $\mathscr{M}_K$ is from 13 % to 37 % of $\mathscr{M}_\Sigma$.

The multi-resolution model in [10] is the first one integrating a geometric and a morphological representations of a scalar field. The main drawback is that it is verbose. As the authors note, an explicit representation of DAG $\mathscr{M}_G$ could be avoided by marking those nodes of $\mathscr{M}_K$, which correspond to the morphological modifications. Moreover, simplification operators applied to Σ may generate new critical points.

Recently, in [19], a new multi-resolution model for triangle meshes Σ endowed with a scalar field has been proposed. The model, rooted in discrete Morse theory, is called Hierarchical Forman Triangulation (*HFT*). It is based on a discrete Morse gradient computed on the given triangle mesh agreeing with the scalar field values known at the vertices of the mesh. The model is then built through interleaved sequences of edge-collapses, reducing the size of the triangle mesh, and of topological simplification operators, reducing the morphological representation. Edge-collapses applied, called gradient-aware edge-contraction, are a subset of all the possible edge-contractions applicable on Σ and avoid deleting or creating critical simplices, thus maintaining the gradient behavior at each update. The *HFT* has a lower storage cost than the mesh at full resolution and provides a high flexibility in adjusting the geometric resolution of the mesh to comply with the topological resolution.

6.3 Summary

Scientific data sets are usually noisy and large. The first issue is due to errors in measurements or simulation that have produced the data, and can be solved through simplification operators. The second issue is a consequence of growing technical and computational capabilities of data acquisition, and can be solved through multi-resolution models.

We have reviewed and compared two simplification operators defined in the framework of morphological representation of scalar fields as Morse and Morse-Smale complexes, namely *cancellation* and *remove* operators. Both operators delete two cells from the Morse complexes, and two vertices from the Morse-Smale complex but, unlike *cancellation*, *remove* operator never increases incidence relation in the Morse complexes, or, equivalently, the number of cells in the Morse-Smale ones. A simplification algorithm based on *remove* and applied on the Morse incidence graph, compared to one based on *cancellation*, produces a Morse incidence graph which is 20 % more compressed, in half the time.

We have reviewed morphological multi-resolution representations for Morse and Morse-Smale complexes. Some of them take also geometric considerations in account. The multi-scale morphological models proposed in [2, 3] are designed for 2D scalar fields. The one in [3] numerically modifies also the scalar field to conform with the changed morphology. The multi-resolution morphological model proposed in [7] is designed for scalar fields in arbitrary dimensions, organizes several representations of the Morse complex in a hierarchy and is capable of supporting the extraction of the representation which best approximates the morphology of the scalar field under application-dependent requirements. The major drawback with this model is the handling of the geometry, which is always that of the original full-resolution mesh even when the resolution of the morphological representation is coarser. Thus, a further issue is designing a model capable of interpreting at different resolution both the topology and the geometry of a scalar field. A first attempt in this direction is the hierarchical model for terrains in [10].

References

1. S. Beucher. Watershed, hierarchical segmentation and waterfall algorithm. In J. Serra and P. Soille, editors, *Mathematical Morphology and its Applications to Image Processing*, volume 2 of *Computational Imaging and Vision*, pages 69–76. Springer, 1994.
2. P.-T. Bremer, H. Edelsbrunner, B. Hamann, and V. Pascucci. A topological hierarchy for functions on triangulated surfaces. *Transactions on Visualization and Computer Graphics*, 10(4):385–396, July/August 2004.
3. P.-T. Bremer, V. Pascucci, and B. Hamann. Maximizing adaptivity in hierarchical topological models. In A.G. Belyaev, A.A. Pasko, and M. Spagnuolo, editors, *Proc. International Conference on Shape Modeling and Applications 2005 (SMI '05)*, pages 300–309, Los Alamitos, California, 2005. IEEE Computer Society Press.
4. P.-T. Bremer, V. Pascucci, and B. Hamann. Maximizing adaptivity in hierarchical topological models using cancellation trees. In Moeller T., Hamann B., and Russell R. D., editors, *Mathematical Foundations of Scientific Visualization, Computer Graphics, and Massive Data Exploration*, pages 1–18. Springer, 2009.
5. L. Čomić. *Operators for Multi-Resolution Morse and Cell Complexes*. PhD thesis, University of Novi Sad – Faculty of Technical Sciences, Serbia, 2014.
6. L. Čomić and L. De Floriani. Dimension-independent simplification and refinement of Morse complexes. *Graphical Models*, 73(5):261–285, September 2011.
7. L. Čomić, L. De Floriani, and F. Iuricich. Dimension-independent multi-resolution Morse complexes. *Computers & Graphics*, 36(5):541–547, 2012.
8. L. Čomić, L. De Floriani, and F. Iuricich. Simplification operators on a dimension-independent graph-based representation of Morse complexes. In C. L. Luengo Hendriks, G. Borgefors, and R. Strand, editors, *ISMM*, volume 7883 of *Lecture Notes in Computer Science*, pages 13–24. Springer, 2013.
9. E. Danovaro, L. De Floriani, P. Magillo, M. M. Mesmoudi, and E. Puppo. Morphology-driven simplification and multiresolution modeling of terrains. In E.Hoel and P.Rigaux, editors, *Proc. ACM GIS 2003 - The 11th International Symposium on Advances in Geographic Information Systems*, pages 63–70. ACM Press, 2003.
10. E. Danovaro, L. De Floriani, P. Magillo, and M. Vitali. Multiresolution Morse triangulations. In G. Elber, A. Fischer, J. Keyser, and M.-S. Kim, editors, *Symposium on Solid and Physical Modeling*, pages 183–188. ACM, 2010.
11. E. Danovaro, L. De Floriani, L. Papaleo, and M. Vitali. A multi-resolution representation for terrain morphology. In *Proc. fourth International Conference on Geographic Information Science - GIScience 2006*, Münster, Germany, September 2006.
12. E. Danovaro, L. De Floriani, and M. Vitali. Multi-resolution Morse-Smale complexes for terrain modeling. In *14th International Conference on Image Analysis and Processing*, Modena, September 10–14 2007.

13. L. De Floriani, P. Magillo, and E. Puppo. VARIANT - processing and visualizing terrains at variable resolution. In *Proc. 5th ACM Workshop on Advances in Geographic Information Systems*, Las Vegas, Nevada, 1997.

14. H. Edelsbrunner, J. Harer, and A. Zomorodian. Hierarchical Morse complexes for piecewise linear 2-manifolds. In *Proc. 17th ACM Symposium on Computational Geometry*, pages 70–79, 2001.

15. H. Edelsbrunner, D. Morozov, and V. Pascucci. Persistence-Sensitive Simplification of Functions on 2-Manifolds. In *SCG'06: Proc. of the 22nd Annual Symposium on Computational Geometry 2006*, 2006.

16. M. Garland and P. S. Heckbert. Surface simplification using quadric error metrics. In *Computer Graphics Proceedings, Annual Conference Series (SIGGRAPH '97), ACM Press*, pages 209–216, 1997.

17. A. Gyulassy, P.-T. Bremer, B. Hamann, and V. Pascucci. Practical considerations in Morse-Smale complex computation. In V. Pascucci, X. Tricoche, H. Hagen, and J. Tierny, editors, *Topological Methods in Data Analysis and Visualization: Theory, Algorithms, and Applications*, Mathematics and Visualization, pages 67–78. Springer Verlag, Heidelberg, 2011.

18. A. Gyulassy, V. Natarajan, V. Pascucci, P.-T. Bremer, and B. Hamann. Topology-based simplification for feature extraction from 3D scalar fields. In *Proc. IEEE Visualization'05*, pages 275–280. ACM Press, 2005.

19. F. Iuricich. *Multi-resolution shape analysis based on discrete Morse decompositions*. PhD thesis, University of Genova – DIBRIS, Italy, 2014.

20. Y. Matsumoto. *An Introduction to Morse Theory*, volume 208 of *Translations of Mathematical Monographs*. American Mathematical Society, 2002.

21. T. Weinkauf, Y. I. Gingold, and O. Sorkine. Topology-based smoothing of 2D scalar fields with c^1-continuity. *Computer Graphics Forum*, 29(3):1221–1230, 2010.

Chapter 7
Experimental Analysis and Comparisons

This chapter presents experimental comparisons of different approaches for morphology computation, and their use in applications.

The different approaches to morphology computation cannot be accurately compared, because they work on scalar fields of different dimensions (such as 2D or 3D functions), they accept different input formats (simplicial models or grids), they assume different properties of input data (such as the absence of flat edges), and/or generate their output in different formats (e.g., by labeling the vertices, or the d-cells of the input model). In Sect. 7.1, we analyze the different output formats and their mutual relations. In Sect. 7.2, we present the metrics used to compare the results of different algorithms. In Sect. 7.3, we compare the watershed and the Forman-based approaches, i.e., the two dimension-independent methods, by presenting results in the 2D and 3D cases from [3]. Here, input data are assumed to be in general position, i.e., no two adjacent vertices have the same field value (there are no flat edges). In Sect. 7.4, we compare all approaches in the 2D case with respect to their feasibility and sensitivity to the presence of flat edges in the input field model by presenting results from [5].

In Sect. 7.5, we present a brief survey of the use of morphological representations in application domains.

7.1 Different Output Formats

Algorithms may compute different structures, such as the ascending or descending Morse complex, or the Morse-Smale complex. For simplicity, we focus on the ascending or descending Morse complex, since they can be easily derived from the Morse-Smale complex in case an algorithm computes this latter. Again for simplicity and brevity, we restrict our discussion to 2D scalar fields described as triangle meshes, and to the computation of the ascending Morse complex.

© The Author(s) 2014
L. Čomić et al., *Morphological Modeling of Terrains and Volume Data*,
SpringerBriefs in Computer Science, DOI 10.1007/978-1-4939-2149-2__7

As discussed in Sect. 2.1.2, an (ascending) Morse complex is computed by different methods in different formats.

Region-based methods classify triangles: each triangle is labeled with the seed minimum of the ascending 2-cell (called, in this approach, *region*) it belongs to. We can easily derive a vertex classification, where vertices whose incident triangles belong to the same region are assigned to that region, and vertices whose incident triangles belong to different regions are labeled as watershed vertices. A complete vertex classification (i.e., without watershed vertices) needs heuristic choices.

Boundary-based methods, in some sense, classify edges: they mark edges which form ascending separatrix lines (if we compute the ascending Morse complex, we do not trace descending separatrix lines), i.e., ridges forming the boundaries between regions. We can easily derive a triangle classification, where each region is defined as a connected component of triangles that are reachable from one another through edge-adjacencies without traversing marked edges. The seed minimum of each region is found as the vertex of minimum field value lying inside the region. Similarly, we can easily derive a vertex classification, where each vertex belonging to a marked edge is classified as watershed. Other vertices are classified as belonging to a certain region, if their incident triangles belong to that region.

Watershed methods produce a vertex classification: each vertex is labeled with the seed minimum or maximum of the region it belongs to. In watershed through simulated immersion, a vertex can also be labeled as watershed. In order to derive a triangle classification, we can define edges, whose two endpoints are watershed vertices, as watershed edges, and let them act like boundaries between regions. Regions are found as the connected components of triangles reachable from one another through edge adjacencies, without traversing such edges. However, this may leave unclassified triangles, i.e., those having all three vertices classified as watershed vertices. Such unclassified triangles are called *watershed triangles*. Watershed triangles are generally very few if the input field does not have flat edges, but they can form large edge-connected components if the input field contains them. Also, watershed methods which do not produce watershed vertices, suffer from the problem of classifying triangles whose three vertices have different labels.

Forman-based methods produce all i-cells of a Morse complex with their topological relations. If we restrict our attention to d-dimensional cells, these are represented by a classification of vertices and of d-cells of the input field model, respectively, in the ascending and in the descending Morse complex. Format conversions are possible, as explained above.

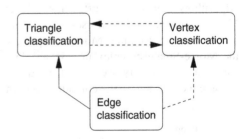

Fig. 7.1 Format conversions for the output of Morse complex computation on a triangle mesh. *Dashed arrows* represent lossy conversion, leaving watershed (i.e., unclassified) vertices or triangles

In summary, given an edge classification, we can consistently derive both a triangle classification and a vertex classification, this latter containing watershed vertices. Given a triangle classification, we can derive consistently a vertex classification, containing watershed vertices.

Given a vertex classification, we can derive a triangle classification, but this may not be complete. If we want a complete classification of all triangles, then we must use some heuristic to assign triangles, for which a unique classification is not implied by the classification of their vertices, to the region of some already classified adjacent triangle. Any heuristic, used to get a complete classification of triangles, is a post-processing algorithm producing new information, not contained in the given vertex classification. In [2] an algorithm for such purpose is presented for triangle and tetrahedral meshes, which can be extended to arbitrary dimensions in a straightforward way. Each d-simplex σ is labeled with the label of the lowest vertex of σ. As d-simplices with all vertices labeled as watershed may exist, they are recursively labeled by propagating labels from $(d-1)$-adjacent d-simplices.

Figure 7.1 shows the various possibilities of format conversion. Note that some of them (represented with dashed arrows) may loose information. For instance, let us start from a triangle classification, get a vertex classification (with watershed vertices), and then go back to a triangle classification. In the last passage, we may introduce watershed triangles. In order to force a complete triangle classification, extra information (like heuristic criteria) are needed, and it is not guaranteed, in general, to obtain exactly the original triangle classification.

Figure 7.2 illustrates the descending Morse complex computed for the knot model (from the AIM-at-SHAPE repository [http://shapes.aimatshape.net/]) with respect to its z-field. Figure 7.2a shows the mesh and z-values, rendered in a color-scale (blue corresponds to low values, red to high values). Figures 7.2b and c show the vertex classification obtained from watershed methods by simulated immersion and rain falling, respectively. Figure 7.2d shows the triangle classification obtained from the region-growing method in [4]. Starting from the vertex classification in (b), the triangle classification in (d) is obtained as, in this specific case, there are no watershed triangles. When starting from the vertex classification in (c), there is a degree of uncertainty in the classification of the strip of triangles having both red and yellow vertices.

7.2 Metrics for Comparison

Assuming that we have two Morse complexes expressed in the same output format, i.e., either a classification of vertices of the input model, or a classification of its triangles, different metrics can be used to evaluate the similarity between them.

We will use three metrics, known in the literature as *Region Number (RN)*, *Rand Index (RI)*, and *Hamming Distance (HD)* [1]. All such metrics assume values between 0 and 1, where 1 corresponds to estimated equality.

Fig. 7.2 (**a**) A triangle mesh and its z-field rendered in a color scale (*from blue to red*). (**b**) Vertex classification from simulated immersion, watershed vertices are in black. (**c**) Vertex classification from rain falling. (**d**) Triangle classification from region-growing. (Color figure online)

RN is rather rough, as it considers just the number of different labels (i.e., the number of regions), present in the two classifications. It is defined as 1 if two classifications have the same number of regions, or as the ratio between the lower number of regions over the higher one, otherwise.

RI measures the probability that a pair of elements has either the same label in both classifications, or has distinct labels in both classifications. It counts the fraction of pairs of triangles both assigned to the same region, or both assigned to different regions. This metric has the advantage of intuitive appeal but becomes less informative as the number of regions grows, going to 1 in the limit as the second term dominates.

HD finds, for each region of one classification, a corresponding region in the other classification, and checks the fraction of triangles assigned to the corresponding regions. It maximizes alignment between two regions and then considers the sum of cardinalities of the symmetric set difference over regions normalized by the total number of classified elements. The complement to 1 of the metric is taken, so that 1 indicates maximum agreement.

In case of vertex classifications having watershed vertices (such as the ones produced by watershed algorithms), the metrics are evaluated considering watershed vertices as neutral.

7.3   Comparing Watershed and Forman-Based Approaches

The watershed and the Forman-based approaches have been experimentally compared in [3]. In particular, the authors have considered the watershed method by simulated immersion [9, 11] and the Forman-based method in [7]. Both methods have been implemented for simplicial complexes of arbitrary dimension with a field value at the vertices.

Such two algorithms have been applied to 2D and 3D scalar fields defined over domain in $\mathbb{R}^2$ and $\mathbb{R}^3$, respectively, and to 2D surfaces embedded in $\mathbb{R}^3$ with an associated curvature value.

All datasets used in the experiments satisfy the theoretical conditions that no two adjacent vertices have the same elevation (i.e., they have no flat edges). For each dataset, the corresponding ascending and descending Morse complexes are extracted using the two approaches and they are compared by using RN, RI and HD metrics.

For the sake of comparison, d-cells of the ascending and descending Morse complexes computed by the Forman-based algorithm are represented as collections of vertices from the input model. The metrics have been evaluated considering watershed vertices as neutral, therefore always considering them as labeled in agreement with their counterpart derived by the Forman extraction.

In Table 7.1, results obtained on terrains (first five datasets), surfaces in 3D (second group of four datasets), and 3D datasets (last six ones) are shown. The two algorithms compute always the same number of ascending and descending d-manifolds, thus, *RN* equals 1. This assures a comparability between the two methods. HD and RI metrics show consistently high similarity values, close to 1, except for the **Neghip** and **Fuel** datasets. Such 3D datasets are characterized by a large volumetric portion of the scalar field occupied by empty space which can generate small discrepancies between the two methods.

Table 7.1 Values of RI and HD metrics computed between the output of watershed and of Forman-based methods, on 2D scalar fields (terrains), 2D shapes with curvature field, and 3D scalar fields

Model	# vert.	RI asc.	RI desc.	HD asc.	HD desc.
Eggs (terrain)	5,751	1.00	0.99	1.00	0.99
St.Helen (terrain)	133,308	0.99	0.97	0.82	0.94
Monviso (terrain)	263,169	0.99	0.99	0.97	0.98
Genova (terrain)	433,174	0.99	0.99	0.93	0.97
Maggiore (terrain)	810,000	0.99	0.99	0.91	0.96
Retinal (shape)	3,643	0.99	0.99	0.99	0.99
Camel (shape)	9,770	0.99	0.99	0.98	0.99
Bumby Torus (shape)	16,815	0.99	0.99	0.97	0.98
Octopus (shape)	16,944	0.99	0.99	0.98	0.98
Analytic1 (3D field)	68,921	0.98	1.00	0.96	1.00
Analytic2 (3D field)	68,921	0.97	0.97	0.96	0.96
BuckyBall (3D field)	32,768	0.99	0.99	0.87	0.88
Fuel (3D field)	262,144	0.99	0.80	0.99	0.86
Neghip (3D field)	262,144	0.79	0.95	0.80	0.77
Silicium (3D field)	113,288	0.94	0.99	0.89	0.90

7.4   Comparing All Approaches in 2D

Five algorithms, which represent all algorithmic approaches, have been compared
in [5]. The used algorithms are: the boundary-based algorithm by Takahashi
et al. [8, 10] (described in Sect. 3.1.1.1), the region-growing method by Magillo
et al. in [4] (described in Sect. 3.2.2), the watershed algorithm based on simulated
immersion by Vincent and Soille [9, 11] (described in Sect. 4.1), the rain falling
watershed algorithm in [6] (described in Sect. 4.3), and the Forman-based algorithm
by Robins et al. [7] (described in Sect. 5.3.4). All algorithms have been implemented
for triangle meshes.

Watershed and Forman-based algorithms label each vertex with the index of a
2-cell, while boundary-based and region-growing algorithms label triangles. For
comparison purposes, the vertex classification provided by watershed and Forman-
based algorithms has been converted to a (complete) triangle classification (see,
Sect. 7.1).

The algorithms have been compared both in presence and in absence of flat
edges. Triangle meshes without flat edges have been obtained from triangle meshes
representing real terrains with flat edges, by applying the method in [5] to remove
flat edges in a morphologically consistent way (see Sect. 3.2.2). Figure 7.3 shows
one of the original triangle meshes (with flat edges).

7.4.1   Comparison on Models Without Flat Edges

For all algorithms, the number of generated ascending and descending 2-cells is the
same. Therefore, RN = 1. However, boundaries between 2-cells may be differently
drawn by each algorithm. Table 7.4 shows triangles classified differently in models
without flat edges. In absence of flat edges, boundary-based and region-growing
algorithms always provide the same output. The outputs of the Forman-based and
rain falling algorithms are also the same. The output of the simulated immersion
algorithm is usually different. There are between 0.5 % and 2.8 % differently-
classified triangles (with just one exception of 6.7 %).

Table 7.3 shows the values of the HD and RI metrics on three triangle meshes
without flat edges. Boundary- and Forman-based algorithms are not shown because
their results are always identical to the region-growing algorithm and to the rain
falling algorithm, respectively. It can be observed that HD $\geq$ 97 % or $\geq$ 93 % (for
the descending and ascending Morse complex, respectively), and RI $\geq$ 99 % or $\geq$
95 % (for the descending and ascending Morse complex, respectively).

7.4.2   Handling Flat Edges

Many algorithms assume that no two vertices connected by an edge have the same
field value, i.e., there are no flat edges.

Fig. 7.3 One of the original terrains with its flat edges and plateaus (in *red*), representing a portion of **Marcy** mountain. (Color figure online)

Table 7.2 Test terrains used in comparison of the results of algorithms in presence and absence of flat edges

Terrain	# vertices	# flat edges
Ustica	1,128	182
Marcy	3,590	466
Elba	1,335	390

The boundary-based and the Forman-based approaches require that there are no flat edges. The boundary-based approach identifies saddles by examining the neighborhood of each vertex. In case of flat edges, a plateau may act as a saddle, but it is not possible to discover it in a local check. Thus, these non-point saddles will not be found and separatrix lines will not be drawn from them. The construction of the Forman gradient in the Forman-based approach assumes that there are no flat edges in the original field. We see no way for extending it in presence of flat edges.

The region-growing approach in its original definition also assumes that there are no flat edges. However, ad-hoc rules can be added for dealing with flat edges as special cases, as done in [4] (see Sect. 3.2.2).

The watershed approach is developed for grey-level images, in which flat edges are common. Algorithms based on such an approach admit flat edges.

The impact of flat edges on the computed Morse complexes has been experimentally estimated on small-size terrains in order to better highlight issues and features (see Table 7.2).

The algorithms have been applied directly to the original triangle mesh with flat edges (by handling flat edges within the algorithm), and to modified meshes without flat edges (obtained from the method in [5], see Sect. 2.4), and the results have been compared. The two watershed algorithms and the region-growing algorithm have been tested, as the boundary-based and Forman-based method cannot be applied in case of flat edges.

Table 7.3 shows the values of HD and RI metrics on three triangle meshes with flat edges. It can be observed that HD $\geq 85\%$ or $\geq 92\%$ (for the descending and ascending Morse complex, respectively), and RI $\geq 97\%$ or $\geq 93\%$ (for the descending and ascending Morse complex, respectively). Compared with the corresponding triangle meshes without flat edges, the absence of flat edges improves similarity from 0.5 to 13.5 percentage points for HD, and from 0.6 to

Table 7.3 Values of Rand Index (RI) and Hamming Distance (HD) metrics on pairs of Morse complexes computed on triangle meshes with flat edges and without flat edges

Terrain	Immersion/Rainfall		Region/Rainfall		Immersion/Region	
	Ascend.	Descend.	Ascend.	Descend.	Ascend.	Descend.
Marcy (flat)	RI 98.82	RI 97.92	RI 98.53	RI 96.94	RI 98.51	RI 98.08
	HD 98.32	HD 95.63	HD 97.56	HD 94.46	HD 98.10	HD 96.67
Marcy (no flat)	RI 99.77	RI 98.53	RI 99.77	RI 98.89	RI 99.98	RI 99.51
	HD 99.48	HD 97.62	HD 99.45	HD 98.07	HD 99.94	HD 98.72
Ustica (flat)	RI 94.34	RI 94.60	RI 91.46	RI 93.73	RI 95.18	RI 97.98
	HD 92.28	HD 92.80	HD 88.30	HD 92.47	HD 92.70	HD 97.06
Ustica (no flat)	RI 98.81	RI 95.48	RI 99.16	RI 95.56	RI 99.14	RI 99.92
	HD 98.15	HD 93.27	HD 98.67	HD 93.46	HD 98.72	HD 99.81
Elba (flat)	RI 93.41	RI 98.93	RI 93.58	RI 97.30	RI 93.41	RI 98.93
	HD 86.27	HD 94.89	HD 84.98	HD 90.94	HD 86.23	HD 94.89
Elba (no flat)	RI 99.39	RI 99.51	RI 99.86	RI 99.84	RI 99.50	RI 99.64
	HD 97.62	HD 97.62	HD 98.71	HD 99.15	HD 98.15	HD 98.23

Table 7.4 Percentage of differently classified triangles in the Morse complexes computed by the various algorithms on terrains with no flat edges

Terrain	Immersion/Rainfall		Region/Rainfall		Immersion/Region	
	Ascend.	Descend.	Ascend.	Descend.	Ascend.	Descend.
Marcy	0.50	2.80	0.55	2.00	0.06	1.30
Ustica	1.80	6.70	1.30	6.50	1.30	0.20
Elba	2.40	2.40	1.30	0.85	1.85	1.70

7.7 percentage points for RI, with greater improvement for the ascending Morse complex. Improved similarity among different approaches suggests more reliability of computed morphological information on models without flat edges (Table 7.4).

7.5 Summary

Sensor technology and simulation methods continue to improve, resulting in datasets of growing resolution and richness, and therefore size. This poses challenges for data analysis. Morphological structures are of interest in a variety of applications, thanks to their ability to capture the structure of a scalar field in a concise and complete way. In case of noisy data, simplification techniques allow eliminating spurious, or less meaningful, features from a morphological representation. Topological methods are used in spatial data analysis and scientific visualization for their ability to extract essential features from data.

Two-dimensional scalar fields with a plane domain are used in geographic information systems to model relief (so-called height fields, or terrains), atmospheric data, etc. In the case of a terrain, the ascending and descending Morse complexes

Fig. 7.4 The 2-cells of the ascending Morse complex (basins) and of the descending one (mountains) for a terrain representing a portion of the **Dolomiti** mountains

are direct representations of mountains and basins. Figure 7.4 shows mountains and basins of a terrain.

Scalar functions defined on a surface embedded in 3D may model phenomenons existing in the surrounding of the shape (such as, for instance, pressure or temperature). Other times, a scalar function is computed from the surface itself to represent some geometric property, such as curvature. In the latter case, Morse complexes allow to describe the shape in a more concise way, such as the collection of convex / concave areas and their mutual adjacency relations. Figure 7.5 shows a segmentation of the shape of a molecule based on its curvature.

Fig. 7.5 The 2-cells of the descending Morse complex for **Retinal** dataset (representing the 3D structure of a retinal molecule) computed from the distortion field of the triangle mesh

Three- and higher-dimensional scalar fields occur in medical applications (such as data from CT and MRI scans), earth sciences, physics and fluid dynamics. The main challenge here is visualization, which is an essential tool for the user to understand the structure of the data.

Fig. 7.6 Segmentation of **Fuel** data set (simulation of fuel injection into a combustion chamber): the original field, and the descending Morse complex of a distortion field computed on the data

Common methods for visualizing scalar fields belong to two broad classes: methods that render iso-surfaces or interval volumes (i.e., intervals between iso-surfaces) extracted from the scalar field, and methods that render directly the entire volume data through color and opacity. Of course, the two methods can be combined together. Methods of the first class display incomplete data, and may be suitable for some applications, such as, for instance, medical data analysis, where the user is generally interested in some types of tissues (e.g., skin, bones, etc.) corresponding to known field values. Methods of the second class produce images which need interpretation from the user and may be sometimes misleading. For example, the used color-map may give more evidence to a less important feature than to a more important one.

Morphological structures, such as Morse complexes and the Morse incidence graph, provide the foundations for a visualization style where, on one hand, the entire data set contributes to the image, and, on the other hand, the elements constituting the real data features are highlighted. Figure 7.6 show a volume data set and its segmentation obtained by means of Morse complexes. Figures 7.7 and 7.8 display the structure of a scalar field directly by showing critical points (saddles, maxima, and minima) and integral curves connecting them as well as the descending Morse segmentations.

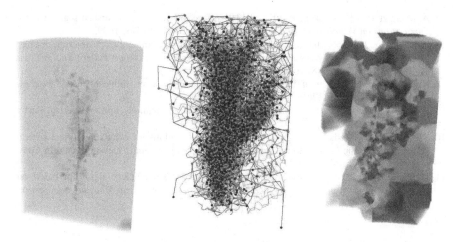

Fig. 7.7 Original field values, 1-skeleton of the Morse-Smale complex and segmentation of the **Fighter** data set, representing data from a wind tunnel model developed at NASA Langley Research Center

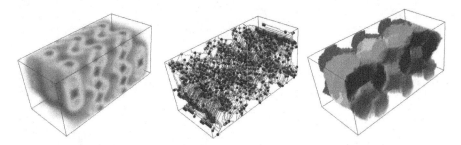

Fig. 7.8 Original field values, 1-skeleton of the Morse-Smale complex and segmentation of the **Silicium** data set, a simulation of a silicium grid

References

1. X. Chen, A. Golovinskiy, and T. Funkhouser. A benchmark for 3D mesh segmentation. *ACM Transactions on Graphics (Proc SIGGRAPH)*, 28(3), 2009.
2. L. Čomić, L. De Floriani, and F. Iuricich. Building morphological representations for 2D and 3D scalar fields. In E. Puppo, A. Brogni, and L. De Floriani, editors, *Eurographics Italian Chapter Conference*, pages 103–110. Eurographics, 2010.
3. L. De Floriani, F. Iuricich, P. Magillo, and P. D. Simari. Discrete Morse versus watershed decompositions of tessellated manifolds. In *ICIAP (2)*, pages 339–348, 2013.
4. P. Magillo, E. Danovaro, L. De Floriani, L. Papaleo, and M. Vitali. A discrete approach to compute terrain morphology. *Computer Vision and Computer Graphics Theory and Applications*, 21:13–26, 2009.
5. P. Magillo, L. De Floriani, and F. Iuricich. Morphologically-aware elimination of flat edges from a tin. In *Proc. 21st ACM SIGSPATIAL International Conference on Advances in Geographic Information Systems (ACM SIGSPATIAL GIS 2013)*, November 5-8 2013.

6. A. Mangan and R. Whitaker. Partitioning 3D surface meshes using watershed segmentation. *Transactions on Visualization and Computer Graphics*, 5(4):308–321, 1999.
7. V. Robins, P. J. Wood, and A. P. Sheppard. Theory and algorithms for constructing discrete Morse complexes from grayscale digital images. *IEEE Transactions on Pattern Analysis and Machine Intelligence*, 33(8):1646–1658, 2011.
8. B. Schneider. Extraction of hierarchical surface networks from bilinear surface patches. *Geographical Analysis*, 37(2):244–263, 2005.
9. P. Soille. *Morphological Image Analysis: Principles and Applications*. Springer-Verlag, Berlin and New York, 2004.
10. S. Takahashi, T. Ikeda, T. L. Kunii, and M. Ueda. Algorithms for extracting correct critical points and constructing topological graphs from discrete geographic elevation data. In *Computer Graphics Forum*, volume 14, pages 181–192, 1995.
11. L. Vincent and P. Soille. Watershed in digital spaces: An efficient algorithm based on immersion simulation. *IEEE Transactions on Pattern Analysis and Machine Intelligence*, 13(6):583–598, 1991.